Diving and Snorkeling Guide to

# The Virgin Islands

DIVING AND SNORKELING GUIDE TO

# The Virgin Islands

by Stephen Bower, Bruce Nyden,
and the editors of Pisces Books.

Pisces Books • New York

**Publishers Note:** At the time of publication of this book, all the information was determined to be as accurate as possible. However, when you use this guide, new construction may have changed land reference points, weather may have altered reef configurations, and some businesses may no longer be functioning. Your assistance in keeping future editions up-to-date will be greatly appreciated.
Also, please pay particular attention to the diver rating system in this book. Know your limits!

Third Printing 1988

 For information about this book write: Pisces Book Co., Inc. (a division of PBC International, Inc.), One School Street, Glen Cove, NY 11542.

**Library of Congress Cataloging in Publication Data**

Bower, Stephen, 1947–
Diving and snorkeling guide to the Virgin Islands.

Bibliography: p.
1. Skin diving—Virgin Islands of the United States—Guide-books. 2. Scuba diving—Virgin Islands of the United States—Guide-books. I. Nyden, Bruce. II. Title.
GV840.S78B635 1984 797.2'3'0972972 84-1063

ISBN 0-86636-032-8

Staff

| | |
|---|---|
| Publisher | **Herb Taylor** |
| Project Director | **Cora Taylor** |
| Series Editor | **Steve Blount** |
| Editors | **Carol Denby** |
| | **Linda Weinraub** |
| Assistant Editor | **Teresa Bonoan** |
| Art Director | **Richard Liu** |
| Artists | **Charlene Sison** |
| | **Alton Cook** |
| | **Dan Kouw** |

Printed in Hong Kong

10 9 8 7 6 5 4 3

# Table of Contents


**How to Use This Guide** 6

**1 Overview of the U.S. Virgin Islands** 8
St. Thomas • St. Croix • St. John

**2 Diving in the U.S. Virgin Islands** 24

**3 Diving in St. Thomas and St. John** 30
Carval Rock • Congo Cay • Capella Island (South Face) • Pinnacle at French Cap Cay • Cow and Calf Rocks • Frech Cap-Cathedral • Saba Island-S.E. Reef • Wreck of the Cartanser Senior • Armado's Reef

**4 Diving in St. Croix** 58
Frederiksted Pier (day) • Frederiksted Pier (night) • East Slope of Salt River Canyon and the Pinnacle • West Wall of Salt River Canyon • Northstar Beach • West End Reef: Butler Bay • Cane Bay Drop-Off • Davis Bay • Buck Island-Scuba Area

**5 Safety** 88

**Appendix 1:** Information and Services 92
**Appendix 2:** Further Reading 93

**Index** 94

# How to Use This Guide

This guide is designed to acquaint you with a variety of dive sites and to provide information that you can use to help you decide whether a particular location is appropriate for your abilities and intended dive plan (e.g., macrophotography, high-speed drift, etc.). The guide also gives a brief history of the Islands, what to look for while sightseeing, where the nightlife is, or where one can go to get away from it all. TheVirgin Islands has it all.

## The Rating System for Divers and Dives

Our suggestions as to the minimum level of expertise required for any given dive should be taken in a conservative sense, keeping in mind the old adage about there being old divers and bold divers but few old bold divers. We consider a *novice* to be someone in decent physical condition, who has recently completed a basic certification diving course, or a certified diver who has not been diving recently or who has no experience in similar waters. We consider an *intermediate* to be a certified diver in excellent physical condition who has been diving actively for at least a year following a basic course, and who has been diving recently in similar waters. We consider an *advanced* diver to be someone who has completed an advanced certification diving course, has been diving recently in similar waters, and is in excellent physical condition. You will have to decide if you are capable of making any particular dive, depending on your level of training, as well as water conditions at the site. Remember that water conditions can change at any time, even during a dive.

*Thick growths of large brain and elkhorn coral colonies are common in the Virgin Islands. Photo: S. Bower.* ▶

# 1

# An Overview of the United States Virgin Islands

The heritage of the United States Virgin Islands, can be plainly seen in its names, faces, architecture, and culture. Many nations have laid claim to these beautiful islands, which are positioned strategically in the northeast corner of the Caribbean Sea; the islands are now a self-governing United States Territory, but the flags of Spain, France, Holland, England, the Knights of Malta, and Denmark have flown over them.

**Early History of the Virgin Islands.** The original inhabitants of the islands, the peaceful Carib tribe, were eradicated by the warrior Arawaks. During his second trip to the New World (1493), Columbus "discovered" and named the islands. Apparently a bit overawed by the number of islands and cays he saw, Columbus named them for St. Ursula and the 11,000 virgins. The land was claimed for Spain and the largest island named Santa Cruz (now St. Croix). The following centuries saw numerous unsuccessful attempts to create a permanent settlement, including three separate tries by the French. These years included what is probably the most flamboyant period in the island's history, as the seedy likes of Bluebeard, Blackbeard, Captain Kidd, and other high-seas pirates secreted their ships and their booty in the many secluded bays of St. Thomas and St. John. In fact, nearby Norman Island is reputed to be the inspiration for Robert Louis Stevenson's *Treasure Island.*

Successful settlement came first to St. Thomas in 1672, when the Danish West India Company established a commercial trading center in Charlotte Amalie. In 1733, the Danes purchased St. Croix, and it quickly prospered as a major sugar producing area. The large flat plains in the center of the island proved ideal for the cultivation of sugar cane, and at one time there were as many as 350 sugar plantations in operation on the island. The ruins of some of the plantation "great houses" still dot the island. The colonists of each nation left their mark and evidence of their stays still can be seen. A map shows a bay named Santa Maria, a region named Bordeaux, a Dronningens Gade street, and a Christiansted city.

In 1916, the United States became concerned about protecting its interests in the Panama Canal, and as major shipping lanes to the Canal pass close to the Virgin Islands, the U.S. recognized the strategic impor-

*Sitting athwart the strategic Sir Francis Drake Channel, the Virgin Islands have been ruled successively by Spain, England, Holland and the United States. Today, their underwater realm belongs to divers, who flock here to view lush coral and marine life.* ▶

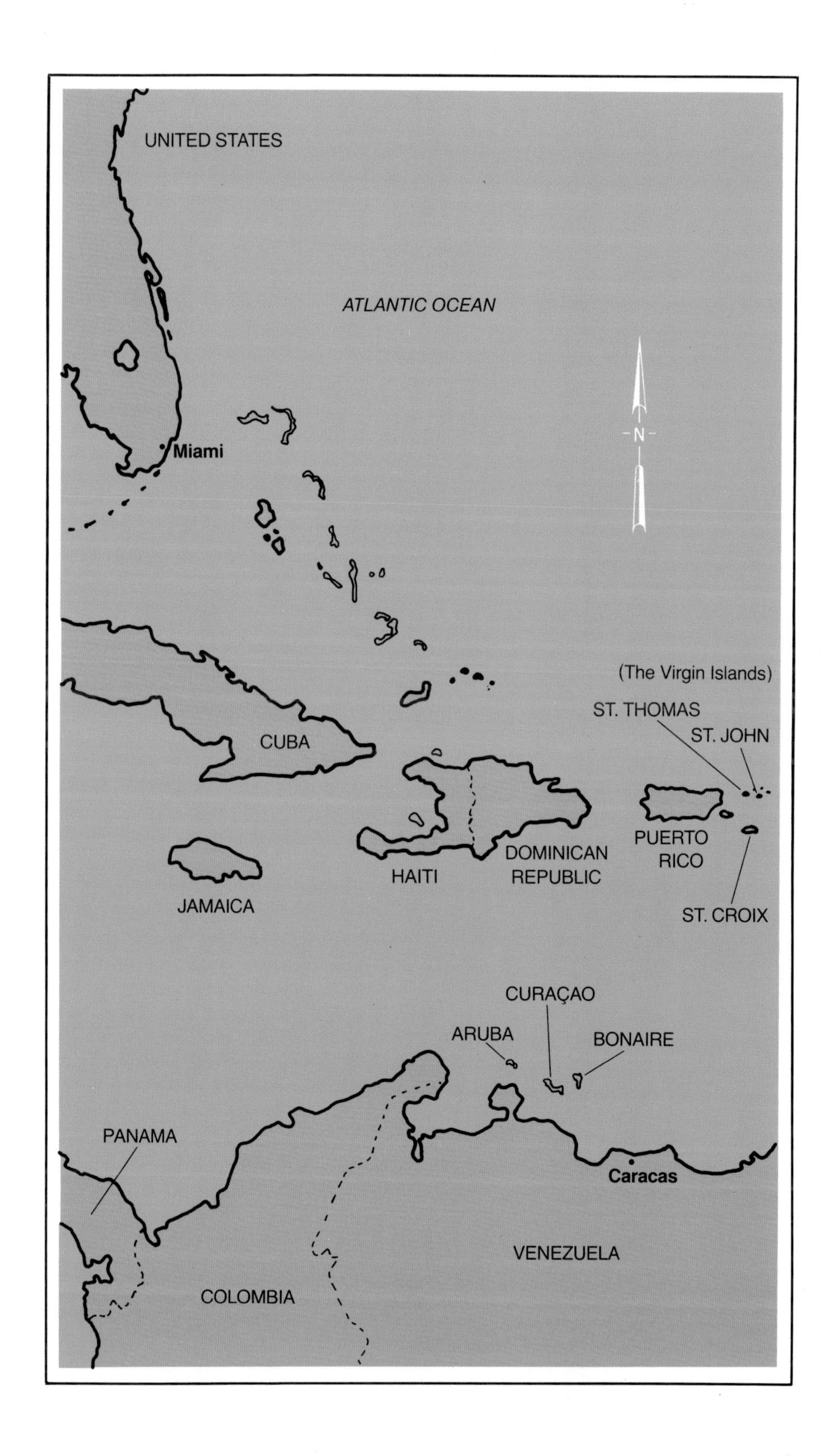
UNITED STATES
ATLANTIC OCEAN
Miami
N
(The Virgin Islands)
ST. THOMAS
ST. JOHN
CUBA
PUERTO
RICO
DOMINICAN
REPUBLIC
HAITI
JAMAICA
ST. CROIX
CURAÇAO
ARUBA
BONAIRE
PANAMA
Caracas
VENEZUELA
COLOMBIA

*The Club Commanche in the Pan Am Pavillion in Christiansted on St. Croix is just one of the cosmopolitan nightspots in the Virgin Islands. Photo: S. Bower.*

*Christiansted's polyglot heritage is evident in its architecture, which is largely Dutch, and its profusion of both English and Spanish place names. Photo: S. Bower.*

tance of these waters. The following year, the Danish islands of St. Thomas, St. Croix, and St. John (along with approximately fifty smaller islands and cays) were purchased by the U.S. for $25 million. For fourteen years, the land remained under the auspices of the Department of the Navy, a sign of the military usefulness of the area. The U.S. Department of the Interior took responsibility for the islands in 1931, and home rule came slowly, not becoming complete until 1970. The dramatic increase in air travel in the 1950s and 1960s brought a tourist boom to the islands, particularly St. Thomas, and has probably changed the economics of the area forever. In 1956, part of the smallest island of the three, St. John, was designated a National Park in order to protect its serene beauty.

**Geography.** The U.S. Virgin Islands lie 40 miles (64 kilometers) east of Puerto Rico and are part of the Antilles Islands chain, which separates the Atlantic Ocean and the Caribbean Sea. St. Croix's Point Udall is the easternmost point of land in the United States. The British Virgin Islands are just east of St. John, across the Sir Francis Drake Channel. St. Thomas, the capital and the busiest of the islands, is volcanic in origin, as is evidenced by its steep hills and deeply indented bays—a geography that provides for spectacular views.

Quiet St. John, just a couple of miles east of St. Thomas, is the smallest of the three islands, measuring 9 by 5 miles (14 by 8 kilometers), two thirds of which is National Park. The geography of St. John is similar to that of St. Thomas, but its mood is much more tranquil. Forty miles (64 kilometers) to the south of these two islands lies the largest Virgin, St. Croix, 28 miles (45 kilometers) long by 7 miles (11 kilometers) wide. Not volcanic in origin, the rolling hills in the northwest slope to a costal plain which has allowed the island its agricultural basis. A tropical rain forest thrives in the northwest, while the eastern end of the island is arid and desert-like.

## St. Thomas

St. Thomas is by far the busiest of the Virgin Islands. Serviced by daily direct flights from New York, Miami, and San Juan, it is truly a tourist mecca, replete with duty-free shopping, sightseeing, a wide variety of restaurants and hotels, and enough nightlife to party away many a vacation. The center of all this activity is Charlotte Amalie (ah-*mahl*-ya), whose harbor can be crowded with cruise ships laden with tourists.

**Shopping.** The town is host to over a hundred shops whose wares include jewelry, watches, perfume, liquor, cigarettes, designer clothing and accessories, cameras, and local crafts, all at "duty-free" prices. (There is, in fact, a customs duty, but it is at a very low rate.) Returning U.S. residents are allowed a duty-free shopping quota of $800, with low duty rates for purchases above the $800 allowance. In addition, up to five cartons of

*Cane Bay offers both sun-and-sand relaxation and an exciting wall dive a short swim from the beach. Photo: S. Bower.*

*Viewed from St. John, the tranquil-looking St. Thomas gives no hint that it's a bustling international cruise port.* Photo: S. Bower.

cigarettes and one gallon of liquor (or one gallon of any liquor plus one fifth of Virgin Islands rum) are allowed. An unlimited number of gifts, if valued at $100 or less, can be sent each day to friends or relatives. Bargains can be found (and remember that you won't be paying local sales tax), but if you have a specific purchase in mind it would be wise to check prices at home first. One guaranteed bargain is rum—at about $1.25 per fifth for Virgin Islands rum to $12 per gallon for Mt. Gay from Barbados, it's hard to resist.

**Sightseeing.** The cultural richness and diversity of St. Thomas and the dramatic natural beauty of its setting make it a splendid island for sightseeing. A stroll up to the fortified, 19th century Bluebeard's Castle affords an excellent view of the beautiful harbor that brought so many influences to the Virgin Islands.

**Transportation.** Seeing the rest of St. Thomas can be done a number of ways: by land, air, or sea. On land, taxis are available for tours of any length and itinerary, including stops for lunch, a swim, or shopping. Usually a bit of local lore and a few stories are thrown in on a cab ride. You may also rent a car, grab a map and remember to drive on the *Left.*

Coral World, near Coki Beach, is an interesting stop for anyone interested in the underwater world. Swimming, running, and snorkeling can be found at any of the wonderful beaches that edge the island. Magen's Beach has been ranked as one of the ten most beautiful in the world, and Mandal, Brewer's, and Coki could be considered winners also. Some of the beaches have restaurants and snorkeling or boat rentals, and some charge a nominal admission fee.

If you want to travel by air, half and whole hour small plane air tours are available and will give you unforgettable views and photos of the island, the sea, and the reefs. From this altitude, the Caribbean takes on an amazing quality, like crystal blue glass, dotted with small boats suspended above the coral reefs.

For the less adventurous underwater explorer, glass-bottom boat tours sail from Charlotte Amalie harbor, as do other boats offering a full day sail to a distant beach or a sunset cruise and cocktails around the harbor and nearby islets. Other day sails also leave from Red Hook, Cowpet's Bay, and Magen's Bay. Ferries offer service to St. John and Tortola (a British Virgin Island).

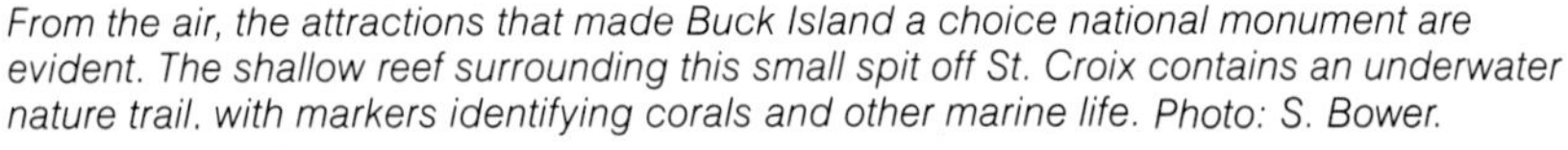

*From the air, the attractions that made Buck Island a choice national monument are evident. The shallow reef surrounding this small spit off St. Croix contains an underwater nature trail, with markers identifying corals and other marine life. Photo: S. Bower.*

**Other Attractions.** Plenty of above-water sports can be found, including tennis, sailing, boardsailing (windsurfing), deep-sea fishing, and golf.

**Dining.** Dining on St. Thomas offers great diversity. While having a meal you can view the sea and sand from a hilltop, or listen to the surf just yards from your table, or sit in a flowered courtyard. You'll have a choice of cuisines from French to Chinese to New York deli. Be sure to try some of the fine, and generally inexpensive, local West Indian restaurants.

As to nightlife, St. Thomas is the Virgin where it's happening if you want it to happen. Disco, steel band, jazz, limbo, a quiet harborside cafe, or a lively hot spot—it's all there.

**Hotels.** In choice of lodgings, the island once more shows its diversity, from a seaside campsite at Hull Bay for very little a night to a suite with a terrace for a great deal, and a lot in between. There is certain to be a place to suit any style and budget. A number of the hotels cater to divers, and staying at one of these will definitely make your dive trip more relaxed and enjoyable. The dive-oriented hotels are listed at the back of this guide.

***Coral World***

If you have friends or family who wonder just what you look at during those hours underwater, take a trip to St. Thomas's Coral World. This is a three-floor underwater observation "tower" sunken onto a coral reef. Windows encircle both underwater levels, and dry divers can take a close look at a few species of well-fed tropical fish, an occasional barracuda or ray, some interesting invertebrates (giant anemone, christmas tree worms, and so on), and some ailing but alive coral formations. Informative wall plaques help identify the most commonly seen marine life. On the grounds near the observatory there is a fine aquarium, a few shops selling ocean-related gifts, and a restaurant-bar with a fine view. Coki Beach is next door, and has good snorkeling, sunning, and swimming. The underwater observatory is open three nights a week, and provides a good look at night-time marine life for those who don't do a lot of night diving.

## St. Croix

St. Croix, the largest of the Virgin Islands, lies to the south of St. Thomas. Many airlines connect the islands, including an exciting and scenic seaplane service between the downtown harbor areas of both islands. All commercial airline flights from the mainland make stops in both St. Thomas and St. Croix.

Although out of the view of most visitors, a few large industries operate on the island, thus giving St. Croix an economic base not entirely dependent on tourism. This makes the island a bit more relaxed than the tourist mecca to the north.

**Christiansted.** Christiansted is quaint, rich in history and Danish architecture, free of the crunch of tourists off the cruise ships, and filled with shops without seeming overly commercial. The largest city on St. Croix, it is still a charming little harbor town on the edge of the Caribbean. The town is centrally located on the north coast of the island, protected from the sea by the two-mile (three-kilometer) Long Reef. Although the urban area sprawls south into the hills, most of the activity and charm of Christiansted is to be found in the older section around the harbor.

A shopping and sightseeing tour of Christiansted can easily be combined on a day's stroll through town. In the 18th century the Danes built Fort Christianvaern to protect their city, and now the fort, from battle-

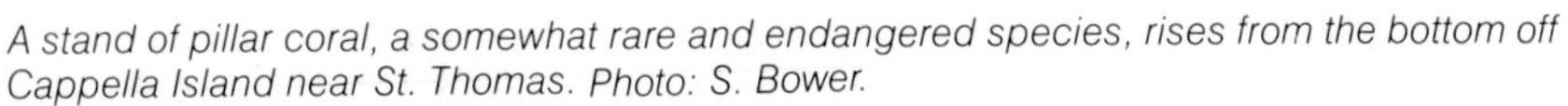

*A stand of pillar coral, a somewhat rare and endangered species, rises from the bottom off Cappella Island near St. Thomas. Photo: S. Bower.*

*Quaint and unusual structures, such as this one in the harbor at Christiansted, are the rule, rather than the exception on St. Croix. Photo: S. Bower.*

ments to dungeons, is open to visitors. The open market on Company Street is the place to shop for local fruits, vegetables, and baked goods. A trip to this market is an experience in the tastes, the accents, and the people of the West Indies.

The restaurants of Christiansted are, for the most part, a reflection of the quiet charm of the town itself: a courtyard setting with live jazz and local seafood, a second-story garden filled with plants, parrots, and international cuisine, an elegant dining room in a restored plantation "great house," or West Indian specialties in what looks like a junk collector's back yard. Unlike Charlotte Amalie, not all these things are happening on the same night. Christiansted just isn't that bustling.

**Frederiksted.** The other town on St. Croix is Frederiksted, on the island's west coast. A 45-minute drive from Christiansted through the middle of the island brings you to this small town stretched out like a toy village along the edge of the sea. Although it is similar in appearance to Christiansted, many find Frederiksted even more lovely with its fanciful lattice-work balconies, covered walks, and stone arches. At the north end of town, the recently restored Fort Frederik is open to the public. (The first salute to the flag of the United States by a foreign power occurred at this fort in 1776.)

**Sightseeing and Entertainment on St. Croix.** There is a lot to St. Croix outside the two towns, and land tours by taxi or bus can be arranged. Rental cars are available through the major chains (offices in the airport and Christiansted) or from smaller local companies, where the cars and the rates are likely to be a bit cheaper but just as reliable. Because St. Croix is a large island, a car is a great help if you want to do some exploring.

**Beaches.** The beaches of St. Croix may not be as spectacular as those of St. Thomas, but there are many beautiful ones. The west end of the island (the Frederiksted end) has Sandy Point, Rainbow, and Sprat Hall beaches, the latter two of which have small bar/restaurants. For a different type of shore try Monk's Baths Beach, which has only a small sand beach but the magnificent coral/rock formations and good snorkeling that more than make up for the skimpy sand. On the north shore, 45 minutes west of Christiansted, Cane Bay and Davis Bay have classic Caribbean beaches—fine white sand, shady palms, clear water, and gentle surf. On the east end, both the Reef Beach at Tague Bay and Cramer Park Beach are beautiful.

Glass-bottom boats make daily tours of the Buck Island Reef National Monument, as do sailing trimarans for snorkeling trips, sunset cruises, or

*Salt River Canyon on St. Croix is a deep ravine that flows out of the shallow bay and spills over a spectacular wall just below the outer line of breakers. Photo: S. Bower.*

*The snorkeling trail around St. Croix's Buck Island attracts numerous visitors. For many, it's the site of their first underwater experience. Photo: B. Nyden.*

all-day excursions to the island's reef and magnificent beach. Also on the west end of the island, the St. George Village Botanical Gardens is a wonderful spot to spend part of an afternoon. Planted amid the ruins of a 19th-century sugar plantation are the terrestrial equivalents of a coral reef. The scores of tropical flowers, plants, and fruit trees are well-labeled and beautifully arranged among the crumbling stone walls. To the northeast of Frederiksted, in the hills, there is a small but lush tropical rain forest which can be seen from the rough, but usually passable, Creque Dam and Annaly Roads.

**Other Above-Water Attractions.** Plenty of non-diving sports are available on St. Croix. Two 18-hole and one 9-hole golf courses, numerous tennis courts, horseback riding, windsurfing and big game fishing are all available to fill up some above-water time.

**Lodgings.** There are many options when it comes to lodgings on St. Croix, including large resort hotels, small amiable hotels, apartments, guest houses, housekeeping cottages, and rental homes. The Virgin Islands Tourist Boards can help with full listings.

## St. John

Descriptions such as tranquil, quiet, unspoiled, peaceful, and low-key are often and aptly used when Virgin Islanders talk of St. John. This, the smallest of the islands, has a terrain similar to that of St. Thomas, with deep secluded bays fringed with white beaches. But there are striking differences between the two islands, and the view from St. John's Mamey Peak or King Hill is one of building free, verdant hillsides and uninhabited islets dotting the sea. St. John has remained so pristine because two thirds of the land and much of the surrounding sea and cays is a U.S. National Park.

Ferries reach the slow-paced town of Cruz Bay after a pleasant 20-minute crossing from Red Hook, St. Thomas. During the day this ferry runs every hour. Other less frequent ferries connect Charlotte Amalie with Cruz Bay, and National Park Dock, St. Thomas, with Caneel Bay Plantation. Land and seaplane flights are available from St. Thomas and St. Croix.

*Caneel Bay, on St. John, is a popular anchorage for cruising yachts. It has both shallow snorkel areas and scuba sites, as well as a hotel along its shores. Photo: S. Bower.*

*St. John's Cinnamon Bay is another popular yacht anchorage. The island contains a national park, and the waters around it are protected.*
*Photo: S. Bower.*

As might be expected, Cruz Bay has a few shops and the restaurants and bars that are requisite on a tropical island near the fast tourist lane, but these establishments have an engaging off-the-beaten-track character that puts the island and the people of St. John apart from their neighbors to the south and west. The ambience of the island attracts celebrities, business people, and professionals, most of whom escape to the high-chic luxury of the Caneel Bay Plantation Rockresort, as well as those on a simpler pilgrimage to the sun and a tent by the beach at Maho or Cinnamon Bay. Between these extremes are guest houses, housekeeping cottages, and rental houses. Those who visit St. John are usually drawn by the beauty and elegance, or by the beauty and simplicity, which makes for an interesting mixture of types.

**Shopping and Entertainment.** Shopping in St. John is fairly limited, but a few merchants in Cruz Bay and the lovely Mongoose Junction sell jewelry, clothing, perfume, liquor, and gifts at the usual duty-free prices. There are no supermarkets on the island, so most people take the ferry to Red Hook to do their marketing.

Most of the island's restaurants and nightspots are in or around Cruz Bay, an area which although never exactly crowded becomes more relaxed in the evening, after the diurnal influx of sightseers have returned to St. Thomas. From West Indian pates and steamed fish in a local cafe to nouvelle cuisine at the Caneel Bay, the selection of restaurants may not be as varied as in St. Thomas, but the island offers plenty of first-rate, charming spots to dine. St. John is not known for its sizzling nightlife, but a few places show old movies or have live entertainment or open-air dancing. The Park Service schedules a number of events daily, including some excellent musical programs in the evenings. A few eateries and pubs are located out of town and provide fine lunches, dinners, and cheerful tranquility. Rental cars and jeeps are available, cabs can be hired for tours or trips, and taxi-buses shuttle to the beaches and the National Park from Cruz Bay. The Visitor's Bureau in town will help with tour arrangements, maps, and information, so try to make use of their service.

*A nest of feather worms blossoms on the surface of a head of star coral. Photo: S. Bower.*

*The vivid colors of marine life seem to almost fluoresce in the gin-clear waters of the Virgin Islands.*

**Sightseeing.** The beaches of St. John are stunning. Trunk Bay is considered one of the world's best, and the Park Service maintains a snorkeling trail with informative underwater plaques, just off the beach. There are also Cinnamon, Maho, and Francis Bay beaches, to mention a few of the larger ones. These are all within the boundaries of the National Park, so they are well-maintained and seldom crowded. The Park Rangers lead nature walks through the lush hillsides of the park, so if you can stay off the reefs for an afternoon and explore the tropical plant and animal life, you may find that St. John's terrestrial life can be almost as interesting as its marine environment. There are a few tennis courts, sailboat and board rentals and instruction are available, a number of boats cater to sport fishermen, and there are hiking trails throughout the Park, so finding sports activities shouldn't be a problem. Day trips to the British Virgin Islands (Tortola, Virgin Gorda, and Peter Island) leave from the Caneel Bay Plantation.

If your want a quiet island with beauty, tranquility, romance, and great diving, you couldn't find a better spot than St. John.

2

# Diving in the U.S. Virgin Islands

There are three large and over fifty smaller, diveable islands and cays with a land area of 135 square miles (337.5 km). More than 200 miles (333 km) of deeply convoluted shoreline encircle the Islands. Twenty-one Dive shops, comprising thirty-four boats visit over sixty of the dive sites, covering a 250 square mile (625 km) area of clear Caribbean water which can be dived 365 days a year. It would take a lot of bottom time to run out of good diving and snorkeling here.

There are many good and practical reasons to consider a dive trip to the U.S. Virgin Islands. Getting there is easy with direct flights from many east coast cities. The major airlines that serve the Virgin Islands offer various discount SuperSaver and APEX fares. Some of these are airfare/hotel packages, and can be most reasonable, particularly from mid-April through November, which is considered "off season."

As the Virgin Islands is a U.S. Territory, many facilities are regulated by U.S. Government agencies, which can make your diving safer and more enjoyable. The dive boats, for instance are registered and inspected by the Coast Guard, and their skippers must be Coast Guard licensed. All the dive guides are instructors, assistant instructors or dive masters certified by American training organizations, and all courses taught are internationally sanctioned and recognized. For the safety of everyone in the water, C-cards and usually logbooks are required by all dive operations, so be sure to have both with you.

Compressors and rental equipment are dependable and maintained at least as well as those on the mainland. Almost all popular brands of diving equipment are available for purchase and can be repaired locally. Of course, U.S. dollars are the legal tender and major credit cards are accepted at many of the dive shops.

In case of emergency, it is reassuring to know that there are complete medical facilities in the Virgin Islands, including; a recompression chamber, only a half hour away in Puerto Rico, another which is frequently available on St. Croix, Coast Guard search and rescue teams, and air evacuation services.

*Divers on St. Croix' Long Reef examine a large basket sponge. Photo: B. Nyden.* ▶

**Weather.** Sea and weather conditions are favorable just about year round. Winter air temperatures average a balmy 82° (28 C) during the day and a comfortable 70° (21 C) at night. The summer temperatures aren't all that much warmer, with averages of 86° (29 C) day, and 75° (23 C) night, plus gentle, but steady Tradewinds provide a cooling breeze, readings above 90° (32 C) are very rare. You can expect it to rain almost every day, that's true, but don't worry, these typical, tropical showers last only a few minutes, and afterward, everything is refreshed and rinsed, and blue skies return. Prevailing winds in the area are controlled by mid-Atlantic weather, that shifts from north-east in the fall to south-east in summer. As these winds shift south, they weaken, and at the same time Atlantic ocean swells diminish, so April to August is a period of calm seas and great visibility. However, late summer is hurricane season. Statistically there are fewer storms in this area than on Long Island Sound, (the last tropical storm to cause any serious damage in the Virgin Islands was in the 1930's.) Mid-August through October tends to be rainier and more humid, and this rain can cause silt run-off, reducing water clarity near shore.

Water temperatures at the surface range from the mid-70's (approx 23 C) in the winter to the lower 80's (approx 28 C) during the summer, but at depth, these gradually drop to about 65° (18 C), so while a wetsuit is not necessary for the occasional dive, some form of thermal protection is in order for anyone spending a lot of time in the water. A wetsuit, or even a pair of dungarees will protect exposed limbs from coral abrasions which can easily become infected by nasties in the marine environment.

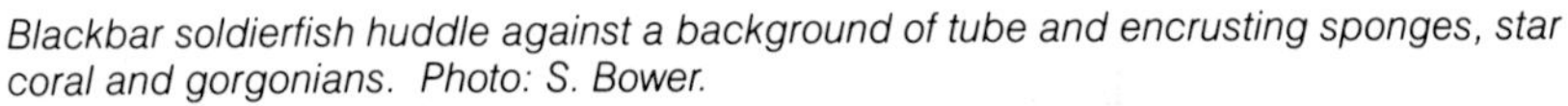

*Blackbar soldierfish huddle against a background of tube and encrusting sponges, star coral and gorgonians. Photo: S. Bower.*

*A pillar coral shows scars where several of its large branching colonies have broken away from the base. Photo: S. Bower.*

**Visibility.** This can be affected by a number of factors; season, weather, sea conditions, distance from land, and other divers. Taking these into account, the Virgin Islands have very good visibility. Averages would be anywhere from 60–120 feet (18–40 meters) with the clearest water surrounding the offshore islands and seamounts. Near the large islands visibility tends to be slightly less due to run-off and wave action in the shallower water, therefore calm, rainless days are usually clearer. The rainy season (late summer through early fall) tends to have more turbid water, and early summer usually the clearest. As might be expected, visibility is often at the mercy of local wind and sea conditions. With so many different dive locations in the area, it's always possible to find one with calm, clear water.

It is very rare to run into a problem with currents in the Virgin Islands. The tidal range is about one foot (.5 meters), so little tidal flow is generated. At times a lot of water does move through the Pillsbury Sound between St. Thomas, and St. John, but protected sites are almost always available. The dive master or guide will let you know of any current, and when beach diving, be sure to find out about the day's conditions from your guide, other divers, or a nearby shop. Remember to begin your dive swimming into the current.

**Marine Life.** Everyone from a novice snorkeler to the experienced diver will find the marine life in the Virgin Islands fascinating and diverse. Large stands of elkhorn and staghorn coral protect the inner reefs, and their inhabitants; peacock flounders, parrottfish, gobies, fairy basslets, damselfish, urchins and anemones. In slightly deeper water stand the larger corals; giant brain coral, tall, fingery pillar coral, branching rose and lettuce-leaf coral. Big vase, basket and colorful tube sponges abound here, and among these are delicate, vivid, angelfish, wrasses, trigger and butterfly fish, to name a few. At the edge of the deep water, wispy gorgonia (soft corals) undulate as a tiny blue chromis, a school of spadefish, or a graceful eagle ray drift by passing a stand of black coral on the reef face. All those brighthued tropical fish from the pages of underwater guidebooks are here, in person, alive and vibrant.

The underwater terrain around St. Thomas and St. John is quite varied, but typical of the area are; outlandish coral formations growing off the rocky substrate, graceful archways of coral harbouring schools of almost metallic glassy sweepers, cavern and tunnel walls vibrant with the colors of sponges, corals and small fish, deep canyons of rock fully coral encrusted and the submerged sides of cays where the sunlight glistens off tarpons schooling near the surface. Beach diving and snorkeling will offer sights of patch coral heads rising from a white sand bottom, a myriad of silversides darting and dancing with the surge, big turtles cruising through the warm shallow water, and stands of elkhorn coral just skimming the shiny surface. There are reef fish of all types as individuals and large healthy schools swim among the coral. A bit offshore, the big pelagic fish; jacks, tarpon, manta and eagle rays, and an occasional shark can be seen.

*The sand flats of St. John's Lameshure Bay are a delight not ony for snorkelers, but for grazing animals such as these star fish. Photo: S. Bower.*

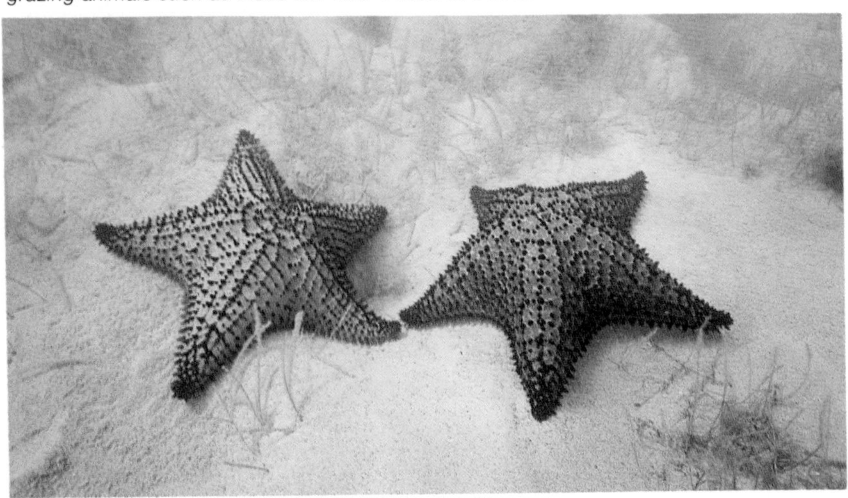

*A diver at the pier in Fredericksted examines a batfish. Photo: S. Bower.*

**Diving Conditions.** Diving conditions in the V.I. are somewhat predictable. Water temperatures range between 75–85 degrees F (25–30 degrees C) year round. Sea conditions are generally 2–3 feet (1 meter) with occasional white–capping. Periodically there are 5–8 foot (2–3 meter) ocean swells generated when a low pressure area passes through, but there are always easily accessible dive spots in the lee of the islands. In the majority of the dive areas listed here, currents are minimal. There are a few places which have one half to 3-knot currents induced by tidal movements. Most dive guides can quickly assess current conditions for visiting divers and will provide alternative locations if conditions are hazardous. Unlike those at many Caribbean destinations, tour operators in the U.S. Virgin Islands must satisfy the same U.S. Coast Guard Safety and licensing requirements as commercial dive operators in the states. You can always be assured that your skipper is experienced and capable.

Beach diving sites are often calm, but many require a fairly long swim out from shore. A few of the beach diving sites on the northern shore of St. Croix characteristically have breaking waves of 2–4 feet (1 meter) in a narrow surf zone. In ideal conditions these sites may have flat seas.

3

# Diving in St. Thomas and St. John

**St. Thomas,** a favored stopover for Caribbean cruise ships, is a diver's island as well. Most dives here are boat dives, with the run to the site ranging from 10 minutes to half an hour. The best sites are generally scuba dives, too deep for most snorkelers to enjoy.

Numerous dive packages, either booked in conjunction with a hotel or with a dive tour operator excluding accomodations, are available. Single tank dives and multi-tank dive trips are available by the day as well.

Instruction and dive services such as gear rental, repair, underwater photo instruction and photo developing are easily available on St. Thomas.

**St. John.** The least commercially developed of the U.S. Virgins, St. John is the site of a national park. Excellent accommodations are available on the island, but it lacks the variety and number of dive operators found on St. Croix and St. Thomas.

There is diving just off the beaches of St. John, and some areas, such as Lameshur Bay, are prime snorkeling territory as well. Lameshur Bay was also the site of the U.S. Navy's Tektite experiments. The Tektite was an underwater habitat that allowed scientists to live underwater indefinitely, giving them more time to study the area's marine life and easier access to the reefs. The project was partially funded by the National Aeronautics and Space Administration (NASA) which was studying the effect of keeping a group of workers together in an isolated environment in very close quarters. While Tektite proved invaluable for marine research, NASA's experiments were unsuccessful. After a few hours at depth, the body becomes saturated with nitrogen, which has a narcotic, and thus calming, effect. The kind of friction between crew members that sometimes develops in space was smoothed over by the slight nitrogen intoxication. Although the Tektite itself has been moved, divers and snorkelers can still visit its location in Lameshur Bay.

▶

*A diver captures a grizzled stand of elkhorn coral on film near St. Thomas. Photo: S. Bower.*

## DIVE SITE RATINGS

| St. Thomas and St. John | Novice Diver | Novice with Instructor or Divemaster | Intermediate Diver | Intermediate with Instructor or Divemaster | Advanced Diver | Advanced with Instructor or Divemaster |
|---|---|---|---|---|---|---|
| 1 Carval Rock | | | | | × | × |
| 2 Congo Cay | × | × | × | × | × | × |
| 3 Capella Island (South Face) | | × | × | × | × | × |
| 4 Pinnacle at French Cap Cay | | | × | × | × | × |
| 5 Cow and Calf Rocks | | | × | × | × | × |
| 6 French Cap-Cathedral | | | × | × | × | × |
| 7 Saba Island — S.E. Reef | | | × | × | × | × |
| 8 Wreck of the Cartanser Senior | × | × | × | × | × | × |
| 9 Armando's Reef | | | × | × | × | × |

*When using the accompanying chart see the information on page 6 for an explanation of the diver rating system and site locations*

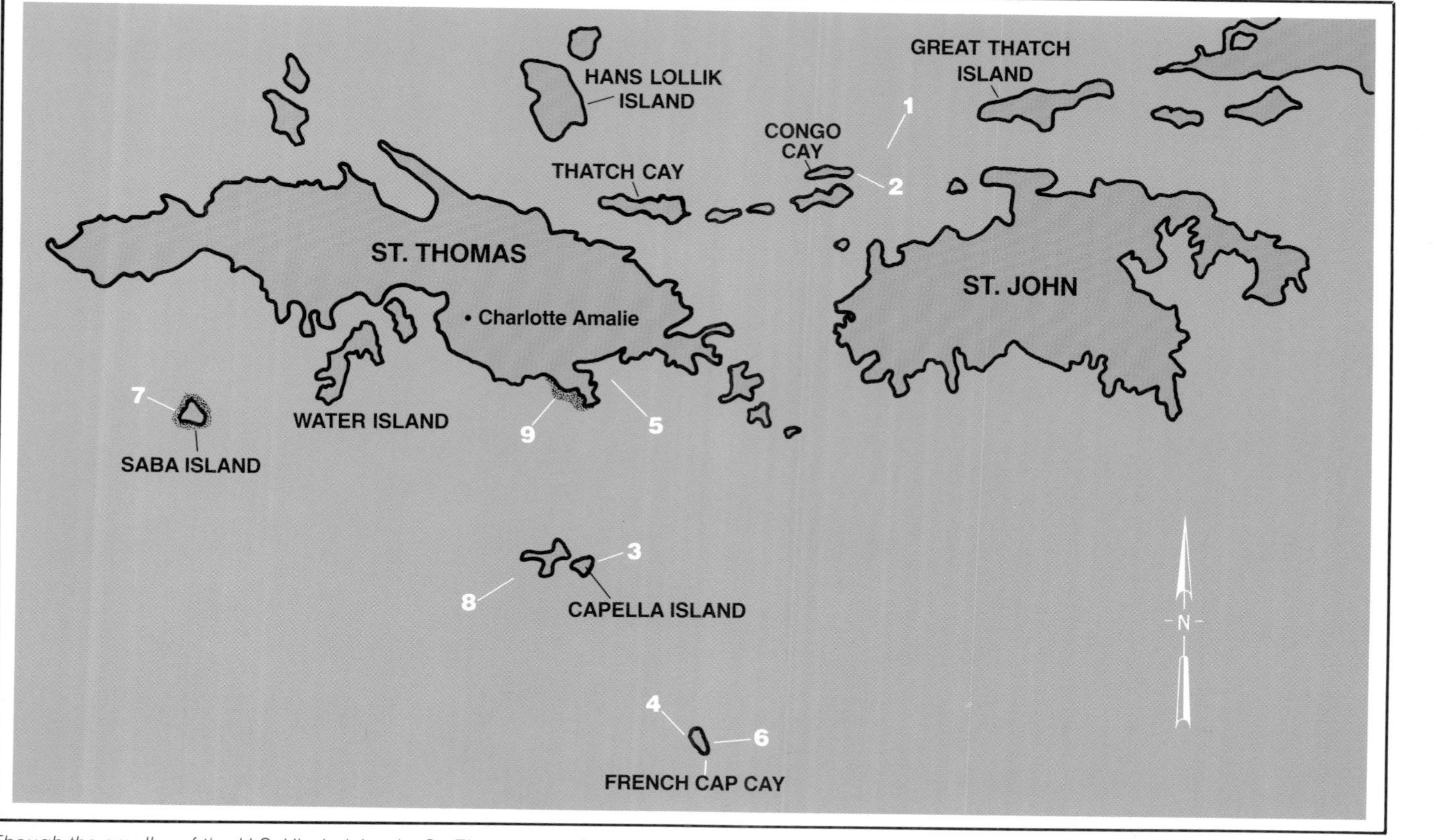


*Though the smaller of the U.S. Virgin Islands, St. Thomas and St. John have no lack of scuba sites for novices or experienced divers.*

## Carval Rock 1

| | | |
|---|---|---|
| **Typical depth range** | : | 20–70 feet (6–21 meters) |
| **Typical current conditions** | : | 1/2 knot common, can exceed 2 knots (unswimmable) |
| **Expertise required** | : | Advanced |
| **Access** | : | Boat |

Carval Rock is hard to class specifically as a St. Thomas dive or a St. John dive because tour operators on both islands run regular trips here when weather permits. The boat ride is approximately the same from St. John as it is from the north side of St. Thomas. Carval Rock is generally classed as an advanced dive because of the strong currents that result from tidal movement through Pillsbury Sound. Occasionally dive guides will make the trip to Carval, inspect the conditions, and then determine whether it is appropriate for the intermediate level. There are a number of excellent alternative sites nearby that are less subject to tidal currents in the event Carval Rock is undiveable.

*Though surge and tidal currents can make Carval Rock a rough dive, the area attracts large marine life, such as manta rays. Photo: S. Bower.*

Carval Rock is the end of a string of stepping-stone cays that cross from St. Thomas to St. John. Dive boats generally anchor on the south side of the rock in 40–50 feet (12–15 meters) of water. The bottom is a dense forest composed primarily of the lavender stalked gorgonian *Pseudopterogorgia.* This particular species of gorgonian is the resting place for basket stars. As you swim out of the gorgonians toward the shallower rock base there appears to be a roadway scratched into the rock. Following this to the southwest you'll pass a narrow cut to the other side of the rock and, a little bit further, a larger one. If surge conditions permit, swim through the larger cut and you'll nearly always find a school (a dozen or more) of tarpon. These fish are 3–4 feet (1 meter) in length and are tame enough to approach almost within touching distance. Clouds of silversides dart and swirl overhead as divers pass through to the other side of the rock. Unless you happen to be diving on an optimal day you should probably not linger

*Large schools of both silversides and tarpon frequent Carval Rock, and can provide brilliant subjects even for photographs using available light only. Photo: S. Bower.*

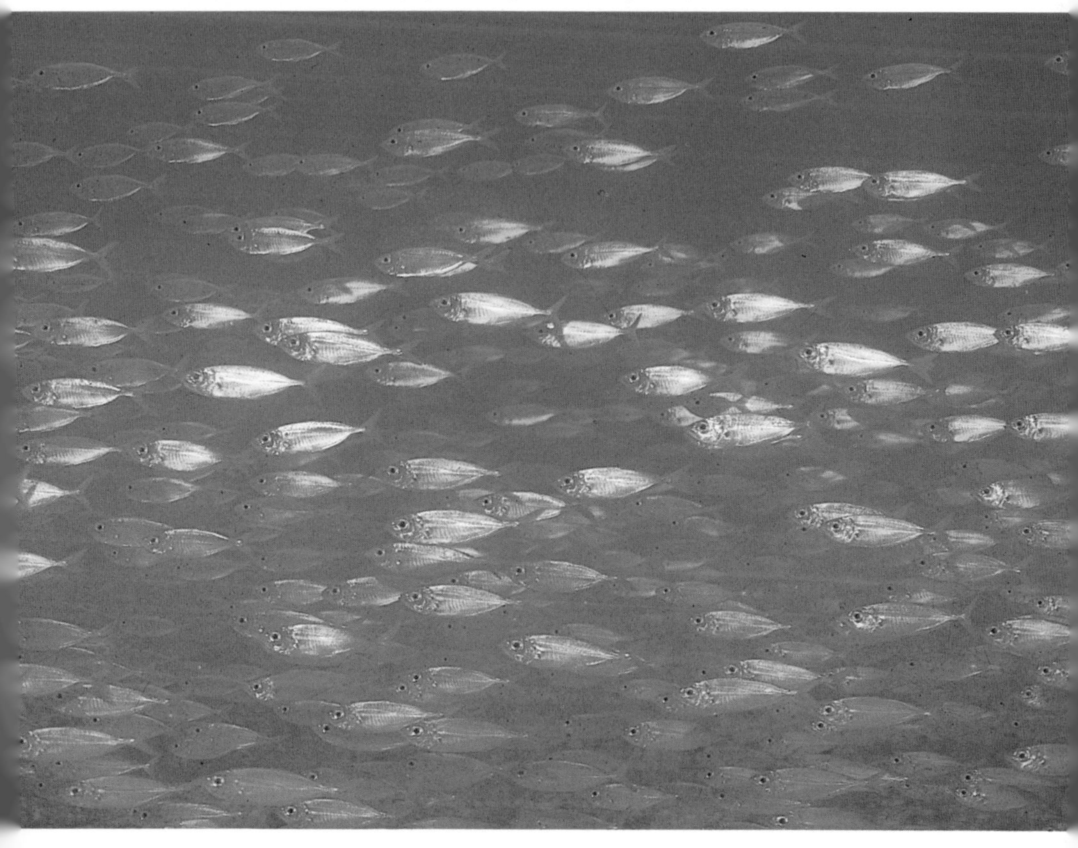

*Lavender-stalked gorgonians are a favored resting place for basket star fish. Photo: S. Bower.*

*The eroded face of Carval Rock juts out of the channel between St. John and St. Thomas, just a few minutes run from St. John. Photo: S. Bower.*

in the cut too long, as the surge can bang you around on the fire coral. On the far side of the rock there are a couple of nice ravines and a sheer vertical face that extends upward to form the north side of Carval Rock. In addition to the tarpon there are some other unique animals to be found at Carval Rock, such as sightings of manta rays. These elusive animals have distinctive scooped mouthparts and can't be mistaken for any other type of ray. To end the dive you can either continue on around the rock and come back down the other side or double back through the cut. Just make sure you have enough air to make it around the end because surface conditions are generally rougher on the windward side of the rock.

**Photo Tips.** The tarpon and the possible manta shots make it worthwhile to set up for large fish portraits. A Nikonos and 28mm will do nicely for the tarpon, and realistically you probably won't be able to get closer than 10 feet (3 meters) to a manta if you happen to spot one, so a moderately wide lens should work out well on both counts. When shooting the tarpon keep in mind that their silvery bodies act just like mirrors and if you broadside them with your strobe you'll end up with a washed spot in your photograph. Try bouncing the light off at a 45-degree angle and stopping down an extra f/stop or two to accommodate for a reflective subject.

## Congo Cay 2

**Typical depth range** : 25–90 feet (8–28 meters)
**Typical current conditions** : Generally less than one half knot
**Expertise required** : Novice (sometimes intermediate)
**Access** : Boat

Congo Cay is one of the string of cays that extends from St. John to St. Thomas. Travel time from St. John is approximately 30 minutes by boat, and the cay is in close proximity to Carval Rock. Dive boats generally anchor in the lee, and because this cay does not protrude into Pillsbury Sound quite as much as Carval Rock, it doesn't seem to have as much tide-related current. The southern end of Congo Cay has a rock ridge that descends into the water and continues on for 225–300 feet (75–100 meters) and disappears in about 90 feet (28 meters) of water. With the boat anchor adjacent to the southern end where the ridge disappears below the wave

*Congo Cay, near Carval Rock, is one of the string of outcroppings in the channel between St. John and St. Thomas. Large growths of corals and gorgonians can be found at its base. Photo: S. Bower.*

*The lush coral outcroppings are typical of the formations at Congo Cay. Photo: S. Bower.*

zone you'll be in 50 feet (15 meters) of water in a nice gorgonian patch. By traveling south around the end and back up the other side you will hit one of the more interesting areas. If you choose to limit your depth you can always cross over the submarine ridge at a shallower point.

Congo Cay is another area where manta and eagle rays are seen frequently, so you should keep a sharp lookout for eerie shadows circling in the distance. Tarpon are fairly common and black tip reef sharks have been observed on occasion. Along the ridge you'll find an interesting array of colorful invertebrates plastered to the rocks. The cup coral, *tubeastrea*, is in abundance on all of the faces. This coral species has flaming orange polyps that remain closed during the day. Occasionally you can find some extended in high-current areas when light levels are low, but you can always count on them delivering a dazzling display on night dives.

To finish off the dive be sure to make at least one pass through the shallow zone where the ridge meets the sea—surge conditions permitting,

*The rock faces at Congo Cay provide an anchor point for many attaching organisms such as this orange sponge. Photo: S. Bower.*

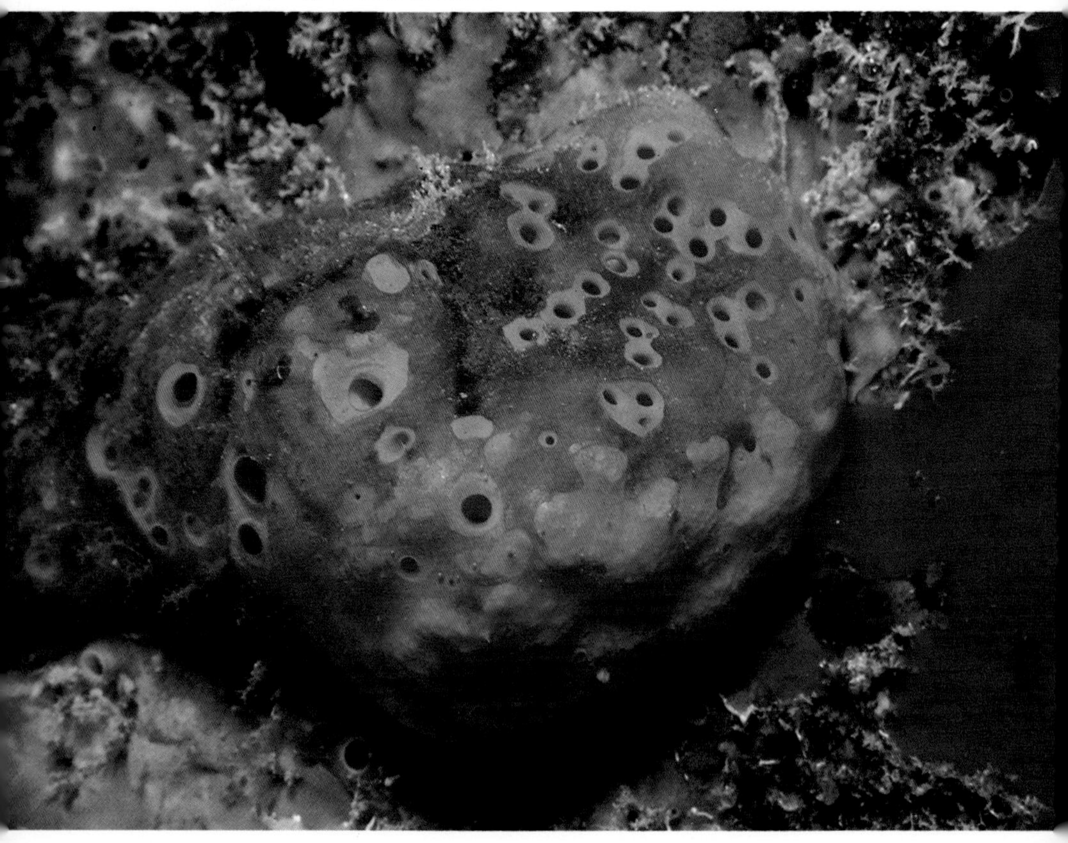

*Brilliant yellow tube sponges and a stand of sea whips attract a photographer's attention. Photo: S. Blount.*

of course. There is a massive cloud of schooling silversides that are said to dwell there nearly year-round. As divers approach the silversides will part and eventually surround the intruders, creating patterns similar to oil and water mixing.

**Photo Tips.** This spot has many of the same photo subjects that Carval Rock offers, and almost any lens you choose will work out well. The school of silversides is probably one of the largest that we've seen in the Virgin Islands and even the available-light photographer will find these to be excellent subjects. These fish seem to prefer the shallower areas, and there is generally enough ambient light to use a shutter speed sufficient to stop the motion of the swirling fish. If your camera system is equipped with a strobe you might want to try shooting some with and some without. The reflective bodies of these fish sometimes cause an otherwise nicely composed shot to be washed out with the reflected light.

## Capella Island (South Face) 3

| | | |
|---|---|---|
| **Typical depth range** | : | 30–70 feet (9–21 meters) |
| **Typical current conditions** | : | Generally none |
| **Expertise required** | : | Novice (under ideal conditions) to intermediate |
| **Access** | : | Boat |
| **Visibility** | : | 10–100 feet (3–30 meters), average 80 feet (26 meters) |

Capella Island is approximately 35 minutes by boat from Charlotte Amalie. It is separated from Buck Island (home of the wreck of the *Cartanser Senior*) by a narrow channel of water and lies slightly to the east of it. This description is for the area below the southeast side of Capella Island. The ocean floor is composed of slanted house-sized blocks of igneous rock, which gives you the impression that you are swimming through giant underwater ravines. Generally the visibility is excellent here and this tends to make the terrain even more impressive. Once again this is an easy boat

*The unusual is usual at Cappella Island, just a short ride from Charlotte Amalie. A unique formation of vase sponges fronts a bed of fire coral. Photo: S. Bower.*

*Like many tropical species, juvenile queen angel fish, while still beautiful, look very different from adults. Photo: B. Nyden.*

entry. With the anchor set on the fringe of this mammoth block reef the boat will generally hang over the nicest portions. If you end up starting the dive near the eastern edge, work your way west along the south side to be assured of covering the best terrain.

Coral growth seems to be limited to the encrusting varieties and several nice stands of pillar coral (that's the stuff that stands upright like furry stalagmites). French and Queen angelfish seem to frequent the area, and for some peculiar reason the blue chromis seem to outnumber the brown by a large margin—unlike similar spots where the reverse is true.

**Photo Tips.** Once again your wide-angle lens will come in handy here. If the visibility is normal you should be able to get some spectacular seascape shots. Try shooting up the face of one the slanted blocks toward the surface, working on the angle that best shows the steepness of the terrain. A hint of flash fill on a colorful subject in the foreground will help to create a really striking shot. If your camera is set up with a portrait-type lens (a Nikonos with 28mm or a 55mm macro on a housed camera are good choices) you can try the old squeeze play on a skittery angelfish.

## The Pinnacle at French Cap Cay 4

| | | |
|---|---|---|
| **Typical depth range** | : | 45–95 feet (14–30 meters) |
| **Typical current conditions** | : | 1/2 knot is common |
| **Expertise required** | : | Intermediate and up |
| **Access** | : | Boat |

The Pinnacle is a small seamount near French Cap's southern point. It is a hop, skip, and a jump from the Cathedral dive, which is located at the opposite end of the cay just 10 minutes away by boat. Since travel time to the cay is generally one hour or more, most tour operators make this a two-tank dive, with one tank done on this deeper spot first and then a second tank at the shallower Cathedral site where the restriction on repetitive dive bottom time will be minimal.

Since there is virtually nothing else around the seamount at depths shallower than 80–90 feet (25–28 meters), most tour operators will try to put the anchor on the Pinnacle itself; this provides a nice guide line to the bottom. The top of the Pinnacle is at a depth of 45 feet (14 meters) and is marked with two stone monoliths. These two rocks have appeared on the

*The two stone monoliths atop the seamount called the Pinnacles have appeared in many published photographs.*
*Photo: S. Bower.*

*Eagle rays can sometimes be seen circling the base of the Pinnacles, and the rock here is covered with filter feeders such as these purple tube sponges. Photo: S. Bower.*

cover of a book and have dressed up the pages of several popular dive magazines. If you tire of watching the schooling fish ride the light current back and forth between the two blocks, there is plenty to see around the base. If you're the first divers in the water you're likely to spot one or more large eagle rays circling the seamount. The 50–80 foot (15–25 meter) range will give you just enough area to cover for a slow scenic tour around the rock base. Visibility is almost always spectacular and the current manageable, but be sure to stay in the general vicinity where you can always find your way back to the boat. If you swim with the current away from the boat you may find yourself treading water for a long time when you surface.

**Photo Tips.** This is an area of superb visibility and from the surface you can usually see the seamount in its entirety. There are fairly nice barrel and tube sponges around. Since the boat is generally hooked on the seamount, you can usually shoot upward with the strobe-lit sponge in the foreground and a clear silhouette of the boat in the distance.

## Cow and Calf Rocks 5

| | | |
|---|---|---|
| **Typical depth range** | : | 25–40 feet (8–12 meters) |
| **Typical current conditions** | : | Generally no current but frequently significant surge action |
| **Expertise required** | : | Intermediate |
| **Access** | : | Boat |

Cow and Calf Rocks are located off the southeast end of St. Thomas, approximately 45 minutes by boat from Charlotte Amalie. The rocks barely protrude above the water's surface, and it is said that the larger two were once spotted by a nearsighted mariner and mistaken for migrating humpback whales, a cow and its calf.

Charter boats generally set anchor at the base of the rock mound, which makes for a convenient entry and guide line to base of the rock where the caves are located. Although surge conditions are generally just a nuisance to divers, the coral-encrusted rocks can give the unwary diver some nasty bruises and scrapes if the shallow areas are approached in rough conditions. The surge is the primary reason this dive is recommended for intermediate divers and up rather than novices.

The most distinguishing feature of this dive site is the coral tunnel network on the west end of the "cow." It branches in a rough H-pattern, is

*Cow and Calf Rocks, though a distance from Charlotte Amalie, are an exciting dive area. A cave at the base of the larger rock is practically silt-free and often houses harmless nurse sharks. Photo: S. Bower.*

plenty wide for two divers and relatively silt free so the end is always in sight. In addition to the tunnel there are at least three major archways, several overhangs, and a few dead-end undercuts. One of the overhangs has a small opening to the surface in the back and is packed with hundreds of glassy sweepers. Schools of large horse-eye jacks frequent the site and nurse sharks occasionally can be seen in some of the caves. This relatively harmless species of shark apparently takes advantage of the water shifting around the base of the rocks to oxygenate their gills in the shelter of the overhangs.

**Photo Tips.** This dive site has some terrific possibilities for wide-angle shots. One favorite trick used by professionals is to back into the opening of a cave (or swim in and turn around if it is wide enough) and shoot toward the opening as their model enters. Try some shots with and without a strobe, using your widest lens. The available light alone will provide some interesting silhouettes. Flash fill can be used to highlight the diver's face and bring out the colorful encrusting marine growth on the tunnel walls. Also, you may be lucky enough to spot a resting nurse shark and the wide-angle lens will do well here too. There is a moderate supply of macro subjects available and a wide variety of local fish species to be found. The coral in general seems to be a little more abused here than in other spots. This is probably due to storm and anchor damage.

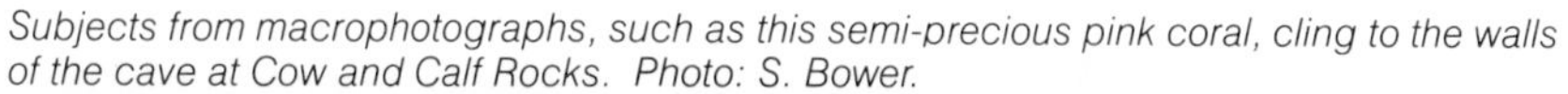

*Subjects from macrophotographs, such as this semi-precious pink coral, cling to the walls of the cave at Cow and Calf Rocks. Photo: S. Bower.*

## French Cap—Cathedral 6

| | | |
|---|---|---|
| **Typical depth range** | : | 15–40 feet (5–12 meters) |
| **Typical current conditions** | : | Less than one half knot |
| **Expertise required** | : | Intermediate |
| **Access** | : | Boat |

French Cap Cay is one of the most distant cays on the south side of St. Thomas. It is approximately one hour by boat from Charlotte Amalie, and depending on the weather can be an uncomfortable ride for those susceptible to motion sickness. French Cap Pinnacle and Cathedral are both superb dives, however, and well worth the trip. This is a description for the shallow shelf on the northeast end of the cay. The shelf has a hard rock bottom with gently rolling contours to the west and deeper ledges to the north. Dive boats generally anchor near the protruding rocks on the

*The Cathedral site at French Cap Cay is a shallow area that is often dived after the deeper Pinnacles site nearby, A large basket sponge protrudes from the sand bottom, which is cobbled with small coral boulders. Photo: S. Bower.*

*A diver examines a basket sponge at French Cap Cay. Photo: S. Blount.*

northwest end of the cay; the Cathedral is located at the base of these rocks in about 20 feet (6 meters) of water. Because of the distance separating French Cap from the mainland, there is very little particulate matter in the water and this accounts for the extraordinary visibility.

Once underwater you'll notice that the rocky substrate is covered with scattered stands of elkhorn coral. Considering the amount of diving that takes place at French Cap, the elkhorn remains relatively unscathed from anchor damage compared to other popular sites. As you approach the base of the protruding rocks you'll notice that it is pocked with coral caves. The easternmost opening is a large archway with a skylight at one end. It is approximately 10 feet (3 meters) wide, 15 feet (5 meters) high, and 50 feet (15 meters) long with a branching tunnel that exits further west. All of the caves and tunnels are spacious, relatively silt free, and navigable by divers.

*No less inspiring than Michelangelo's works in the Sistine Chapel, a mottled mosaic of sponge and algae grace the ceiling of the Cathedral. Photo: S. Bower.*

*The colors of the marine life outside of the cave at the Cathedral are striking. Yellow tube sponges hug a small coral mound. Photo: S. Blount.*

The interior of the cathedral is lined with numerous invertebrates such as encrusting sponges, flame scallops, *tubeastrea* (cup corals), and the small stylaster (2–3 inches or 1 centimeter tall lavender coral with blanched white tips). In the open areas surrounding French Cap Cay there can always be surprises for the avid fishwatcher. Because of the remoteness of the cay you'll be more likely to spot the more elusive types that avoid human contact. We observed schools of a small barracuda-like fish called sennet and some large jacks perusing the area.

This location seems to be one of the favorites in St. Thomas despite the travel time to the site, and we recommend it to all divers. Don't let the shallow depths fool you into thinking this is just for intermediate divers.

## Saba Island (Southeast Reef) 7

| | | |
|---|---|---|
| **Typical depth range** | : | 20–50 feet (6–15 meters) |
| **Typical current conditions** | : | None, some surge action with weather |
| **Expertise required** | : | Intermediate |
| **Access** | : | Boat |

Saba Island is located southwest of Charlotte Amalie and is approximately 25 minutes away by boat. Because of its close proximity to the mainland it is a popular spot for one-tank trips. The underwater terrain on the southeast reef resembles that of Capella Island on a smaller scale. The bottom seems to be strewn with large blocks of rock, and there is one very nice archway in 30–40 feet (10–12 meters) of water.

The Creole wrasse is one of the larger species of wrasse found here in large quantities. They are indigo blue with curious splotches of color arranged in patchwork fashion about the front halves of their bodies. On night dives you can find them bedded down in small cracks and crevices

*Saba Island, close to Charlotte Amalie, has plenty of semi-tame fish. Here a diver holds a filefish. Photo: S. Bower.*

and can approach within touching distance. This area also seems to have plenty of rock beauties and angelfish. There are some nice stands of pillar coral and some small patches of staghorn coral. The staghorn seems to prefer the surge action on the faces of the sloping rocks in the shallower areas.

You might want to bring a pair of old jeans along if it is a particularly surgy day. This spot is heavy with fire coral and when the surge is rough it's pretty easy to brush up against the fire coral if you're not paying attention. This mustard brown encrusting coral grows in several forms and also can take on the shape of an encrusted sea fan or gorgonian. The sting is mild in small doses, but if you get rolled around in it you will probably smart for an hour or so and have some nice welts for a day or two. It's interesting to note that the stinging apparatus, the nematocysts, are unable to penetrate the calloused portions on your palms and finger tips. Most people can touch the coral with the working side of their hands and not even know it.

*A cluster of tube worms burst from their hiding places on the top of a head of star coral. Photo: S. Bower.*

## The Wreck of the *Cartanser Senior* 8

| | | |
|---|---|---|
| **Typical depth range** | : | 25–35 feet (8–11 meters) |
| **Typical current conditions** | : | None |
| **Expertise required** | : | Novice |
| **Access** | : | Boat |

Just off the western end of St. Thomas's Buck Island and less than a 30-minute trip from most of the dive shops lies the wreck of the *Cartanser Senior.* The hull sits in 35 feet (11 meters) of water in a small cove protected from heavy currents and surge, thus making this an easy wreck dive. This site is popular enough to require four permanent moorings to handle busy days. Be sure to note where your dive boat is in relation to the *Cartanser* to avoid the embarrassment of surfacing at the wrong boat. Trips to the wreck are usually made as afternoon single-tank dives, but sometimes this site is visited as the second of a morning departure two-tank trip.

The history of the *Cartanser Senior* is a bit difficult to get straight, but the story has it that the ship was involved in some quasi-legal activity (such as smuggling) and was abandoned by the skipper and crew. Left unmanned and slowly taking on water, it eventually sank in Gregerie

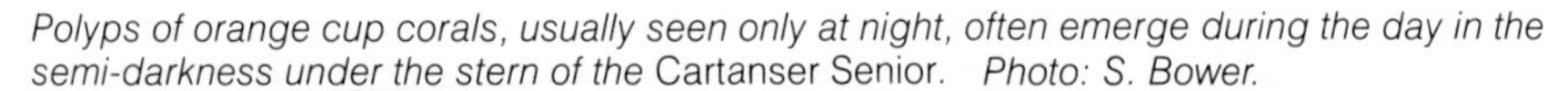

*Polyps of orange cup corals, usually seen only at night, often emerge during the day in the semi-darkness under the stern of the* Cartanser Senior. *Photo: S. Bower.*

*Sergeant majors and a variety of other species now man the empty companionways and cargo holds of the* Cartanser Senior. *Photo: S. Bower.*

Channel in St. Thomas harbor. Deeming it a hazard to navigation, the Corps of Engineers intended to demolish the ship, but their plans were thwarted by an underwater picket line of divers intent on saving their dive site. The situation was resolved by moving the *Cartanser* to a spot off Buck Island in 70 feet (21 meters) of water. A couple of years later, at the insistance of hurricane Alan, the wreck was moved again, this time to its present location.

The *Cartanser* is a good-looking wreck. Having slipped below the waves so gently, it is almost intact except for some slight havoc wreaked by Alan. The ship lies on its port side at a 45° angle on a flat sand bottom. Visibility in the area is generally quite good, so from midships the entire hull can be seen. The holds of the ship are uncovered and many of the interior bulkhead hatches are gone, so you can safely swim through the cargo areas, engine room, and other parts of the ship's interior. Sergeant majors, parrotfish, angelfish, and yellowtail snappers have taken a liking to the wreck, and a variety of encrusting sponges and corals have grown over the steel framework. Numerous clumps of colorful orange cup corals thrive in the relative darkness under the flare at the stern of the ship, and the polyps of these corals, usually seen only at night, are often exposed during the day.

The *Cartanser*, while not as spectacular and steeped in history as the *Rhone*, is a fine wreck dive and is excellent for the first-time wreck diver.

## Armando's Reef 9

| | | |
|---|---|---|
| **Typical depth range** | : | 20–50 feet (6–15 meters) |
| **Typical current conditions** | : | Little to moderate, with occasional surge |
| **Expertise required** | : | Intermediate |
| **Access** | : | Boat |

Armando's Reef is just a quarter of a mile (two fifths of a kilometer) off Cocolus Point, and only a ten minute boat ride from Charlotte Amalie or a number of nearby hotels and dive operations. The reef is named for Armando Jenik, who for many years ran dive operations in the Virgin Islands and can be credited with discovering scores of spots now visited by divers.

The substructure of this reef is a large stone outcropping rising out of the sandy bottom in 50 feet (15 meters) of water. Most of the rock is so fully encrusted that it looks like solid coral, but an occasional, rocky, sheer-sided crevasse divulges its underlying origins. This rock base makes for very steep walls, so the dive here is, in some areas, almost a "mini-wall" dive. The sea fan and gorgonia-covered canyons make for beautiful reef architecture. The walls are covered with red and yellow boring sponges, fire sponge, christmas tree and fan worms, and spiny oysters (so covered with sponges as to be almost invisible until, upon your approach, they snap shut). Every so often a ringed anemone can be found hidden behind a sponge, and nearby, a couple of cleaner shrimp will be waiting for a host in need of a scouring. The tiny marine life could keep you fully interested for

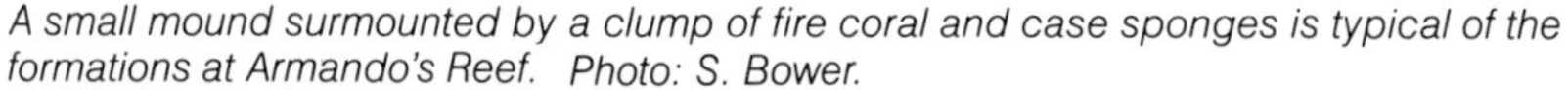

*A small mound surmounted by a clump of fire coral and case sponges is typical of the formations at Armando's Reef. Photo: S. Bower.*

*Squirrelfish and other nocturnal species can be found in profusion in the underhangs at Armando's Reef. Photo: S. Bower.*

a whole dive, the only limiting factor being the often-found surge, which tends to make it difficult to stay in one spot long enough to investigate fully.

The sides of some of the surge channels that run between the coral walls have been undercut at their bases due to erosion by sand as it is swept through by wave action. The shallow caverns thus created provide a home for many types of nocturnal fish who prefer enclosed protection to life on the open reef. These spots are thick with squirrelfish and bigeyes, many swimming upside-down as they orient themselves to the sides and top of the cavern, not to up and down as we know it. Try just watching from outside one of these overhangs as they swim from ceiling to wall to floor, all the time "correcting" their orientation as if to the bottom. Also dwelling in the undercuts are all sorts of grunts, Spanish, French, and black. In the open sandy area between these coral walls can be found many of the common reef fishes. Including the usual damsels, sergeant majors, fairy basslets, chromis, and parrotfish, plus a few large queen parrotfish sometimes referred to as the "super males."

Armando's Reef is a lovely dive site, and because it is close to the dive shops and fairly shallow, it can be visited easily and often.

**Photo Tips.** If heavy surge doesn't make it too difficult, this is a great spot to do some fish portraits. On a day without surge, a macro lens in a housing would be your best bet, and if it is surgy, try a 28mm on the Nikonos for your fish shots. Wide angle work may be hampered by the less than spectacular visibility sometimes found here, but on a good day there is a lot of nice reef configuration to shoot.

4

# Diving in St. Croix

Of the U.S. Virgin Islands, St. Croix is blessed with the most abundant near shore and beach diving. The pier at Fredericksted is acknowledged as one of the leading pier dive sites in the entire Caribbean. Cane Bay, Davis Bay and other areas around the island are easily reached by swimming out from the beach. Gear rental, guides and instruction service are available both for this kind of casual, do it yourself diving and for boat diving to the many spectacular reef areas nearby.

Like St. John, St. Croix has been the focus of numerous marine science experiments. In Salt River Canyon, just outside of Salt River Bay, sits the Hydrolab, an underwater habitat maintained by the National Oceanic and Atmospheric Administration (NOAA). Teams of marine scientists live in Hydrolab, usually for a week to 10 days, studying the geologically-active Salt River ecosystem. When there are no scientists in residence, local tour operators often take divers down to view the habitat.

As with St. Thomas, numerous excellent dive services are available. Packages including accommodations and diving can be arranged with a variety of hotels, or divers can book by the day on dive boats working out of Christiansted.

St. Croix also has Buck Island, one of a handful of national marine parks. Reserved for snorkelers, an underwater nature trail with markers identifying coral and other marine life leads visitors through a lush, shallow reef system. Parts of Buck Island can be dived with scuba as well.

*A large basket sponge and a mound of star coral border the dropoff at Cane Bay on St. Croix. Photo: S. Blount.*

## Dive Site Ratings

| St. Croix | Novice Diver | Novice with Instructor or Divemaster | Intermediate Diver | Intermediate with Instructor or Divemaster | Advanced Diver | Advanced with Instructor or Divemaster |
|---|---|---|---|---|---|---|
| 1 Frederiksted Pier°(day) | × | × | × | × | × | × |
| 2 Frederiksted Pier°(night) | × | × | × | × | × | × |
| 3 East Slope of Salt River Canyon and the Pinnacle | | | × | × | × | × |
| 4 West Wall of Salt River Canyon | | | × | × | × | × |
| 5 Northstar Beach* | | | × | × | × | × |
| 6 West End Reef: Butler Bay | × | × | × | × | × | × |
| 7 Cane Bay Drop-Off | | × | × | × | × | × |
| 8 Davis Bay | | | | | × | × |
| 9 Buck Island-Scuba Area* | × | × | × | × | × | × |

* Indicates Good Snorkeling Spot

° Good Night Diving for Intermediate level or better

*When using the accompanying chart see the information on page 6 for an explanation of the diver rating system and site locations*

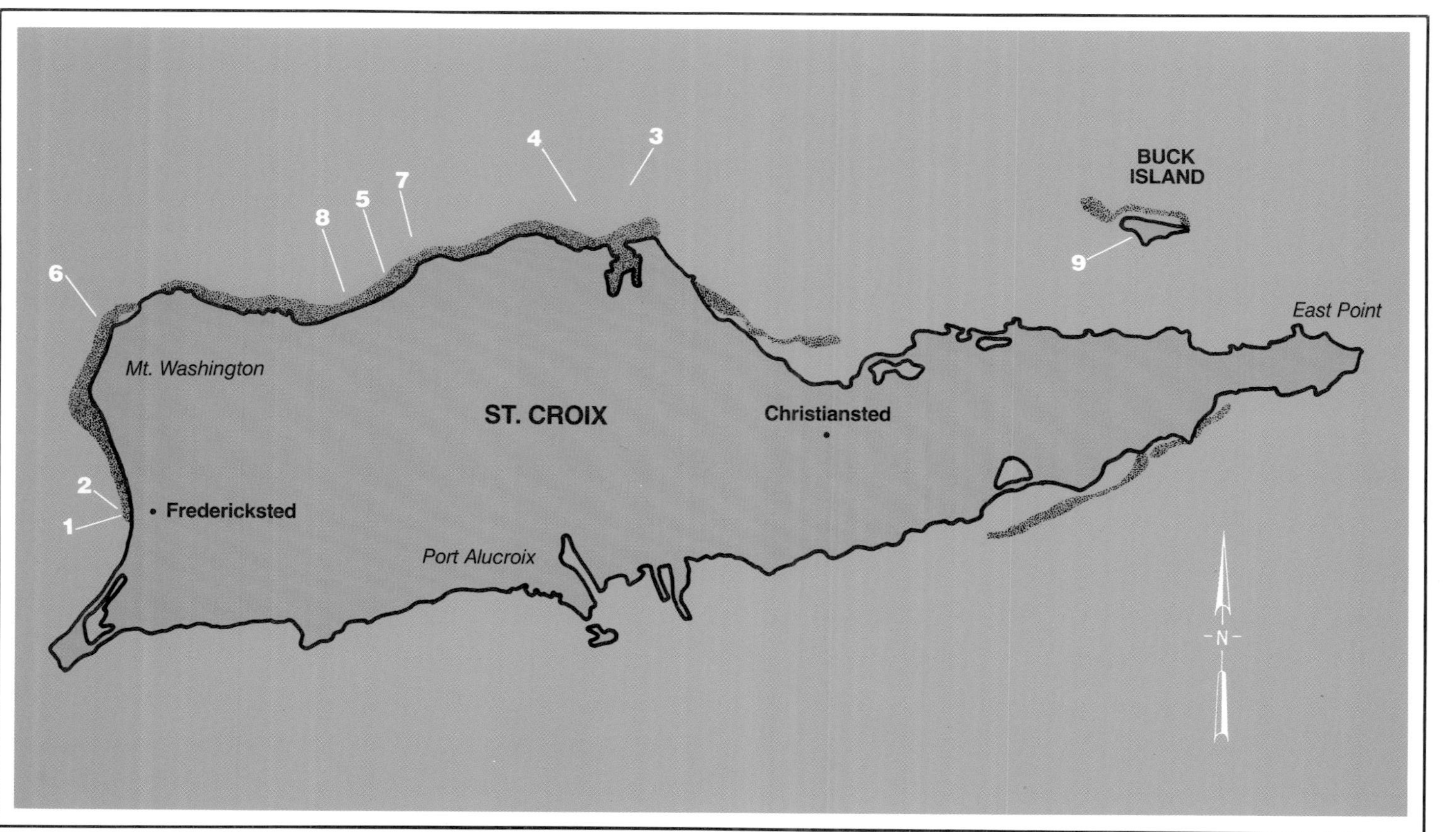


*St. Croix may be the most "American" of the U.S.V.I.s. It also has a greater number of shore dives than its sisters and a variety of topside attractions.*

## Frederiksted Pier (Day) 1

| | | |
|---|---|---|
| **Typical depth range** | : | 10–35 feet (3–11 meters) |
| **Typical current conditions** | : | None to light |
| **Expertise required** | : | Novice |
| **Access** | : | Entry is off the pier itself (exit up a ladder onto the pier) |

One of the most unusual and interesting dives on St. Croix is under the pier in the town of Frederiksted. This large and fairly busy commercial pier runs 450 yards (about one half kilometer) out from the main street in the town which is St. Croix's second largest city. Since its construction, the pilings supporting the pier have been overgrown with encrusting marine life, and provide a haven for small free-swimming species.

Most of the dive shops run well-organized day and night dives at the pier, but if you are doing it yourself, here are a few tips: The harbor master enforces the sensible rule of no diving when there are commercial or naval ships at the pier, so it is recommended that you call the day before *and* the

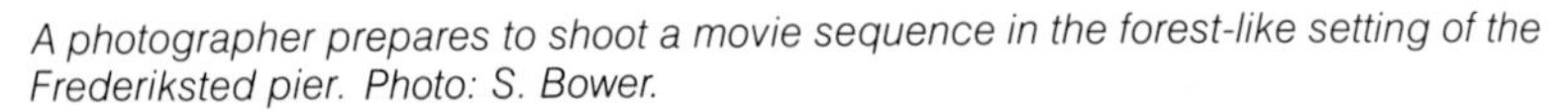

*A photographer prepares to shoot a movie sequence in the forest-like setting of the Frederiksted pier. Photo: S. Bower.*

*The pilings of the Frederiksted pier are a visual riot of color. A juvenile French angelfish noses among yellow tube sponges. Photo: B. Nyden.*

day you plan to dive, as ships can rule out diving for days in a row. The harbor master's number is 772-0174. When you arrive at the pier, stop at the office. You may need to sign a release form and show your C-card. The best access to the water is at the widened portion of the pier, where a steel ladder descends the 15 feet (5 meters) into the water. The easiest procedure is to gear up fully from your car, then just make a stride entry off the edge of the pier. If an entry from this height is terrifying, you can use the ladder.

At the ladder there is about 20 feet (6 meters) of water, with depths varying from 35 feet (11 meters) at the outer end to 10 feet (3 meters) toward shore. The pilings are covered with marine life from the surface to the bottom and you can dive at any depth, but staying near the sandy floor is usually easiest and most interesting. The outer and wider portion of the pier has many more pilings to investigate, so this is probably the best direction to head, but there is plenty to see inshore as well.

*Certainly among the ugliest of ocean species, batfish are especially prevalent at the Frederiksted pier. Photo: B. Nyden.*

### *Bottle Diving*

The U.S. Virgin Islands have been visited repeatedly by mariners since Christopher Columbus first landed in St. Croix in 1493. Sailors have always tossed their trash overboard, and areas of high activity such as Charlotte Amalie, Christiansted,and Frederiksted harbors are littered with bottles. Although it is illegal to disturb a historical wreck site or to dive in a navigable waterway, solitary bottles and artifacts can often be found by poking around the murky harbor areas. One collection of these pieces, the John Damron collection, spans three and a half centuries, beginning with a Spanish amphora dated 1625.

The marine life here is bountiful to say the least. Sponges have adhered to the pilings: orange and brown tube, red boring, red fire (caution) and red finger, brown touch-me-not, iridescent tube, and gray cornucopia. Each sponge is home for at least one arrow crab or brittle star, and fire worms (again, caution) inch across their surface. Encrusting corals of many colors also decorate the pilings. Christmas tree, fan, and feather duster worms are everywhere and sea urchins are hiding wherever they can find a hollow. During the day, the seahorses are so well camouflaged that they are very difficult to spot, but they are there. The fish population is greater and more diverse at the outer and deeper end of the pier. Here you will find schools of snapper, grunt, and sergeant major, along with trunkfish, burrfish (a puffer), trumpetfish, goatfish, and an occasional stingray. A number of quite unusual species inhabit the pier, like the long-lure frogfish, whose camouflage makes him almost indistinguishable from a piece of sponge, or the other master of disguise, the stonefish, with its poisonous dorsal fin. You might also see a long-nosed batfish walking the bottom on its pectoral fins, or a snake eel's fat head sticking out of the sand, or a pair of gurnards "flying" over the sand.

During the day the pier is a great and easy dive, and only the entry and ladder exit might cause any difficulty for the novice diver.

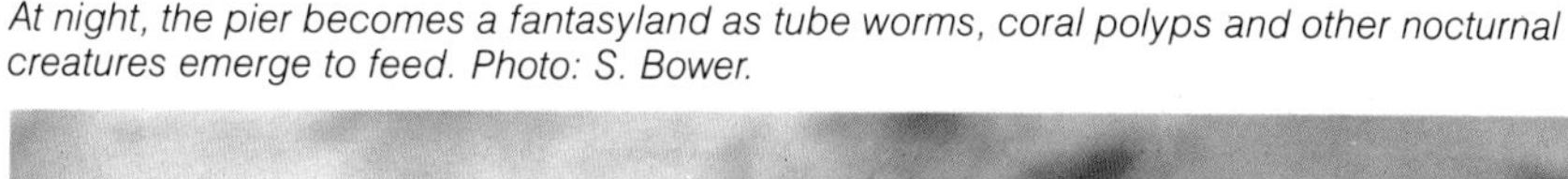
*At night, the pier becomes a fantasyland as tube worms, coral polyps and other nocturnal creatures emerge to feed. Photo: S. Bower.*

## Frederiksted Pier (Night) 2

| | | |
|---|---|---|
| **Typical depth range** | : | 0–30 feet (10 meters) |
| **Typical current conditions** | : | None to very light |
| **Expertise required** | : | Intermediate |
| **Access** | : | Directly off pier |

During the day the pier in Frederiksted is a fascinating dive; at night it's a spectacular dive. Try to make a dive in daylight prior to your first night dive at the pier. The contrast between the two makes the whole experience more rewarding, and at night the pier takes on a wonderful eerie quality that is easier to appreciate if you have seen it during the day.

Entry and exit techniques are similar to those of day diving on the pier, but a few precautions should be taken. Schools of jellyfish (*cubo medusae* or sea wasps being the most common and nastiest) occasionally drift through the area at night, so look with your dive light before you leap off the edge to be sure that the water is clear of the creatures, and try not to linger at the surface on either entry or exit as they tend to float just below the surface. Before you enter check to be sure there is no one below you, as a diver without a light is very hard to see in the water at night. Once in the water you can tie a cyalume light stick to the bottom of the ladder to help you find your way back, and to keep you oriented during the dive.

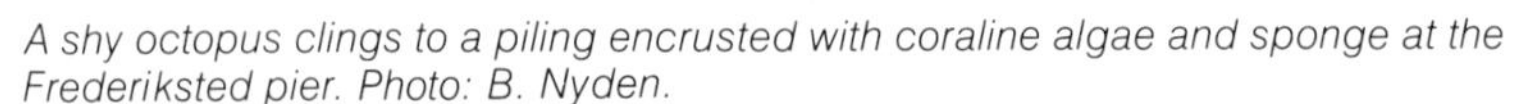

*A shy octopus clings to a piling encrusted with coraline algae and sponge at the Frederiksted pier. Photo: B. Nyden.*

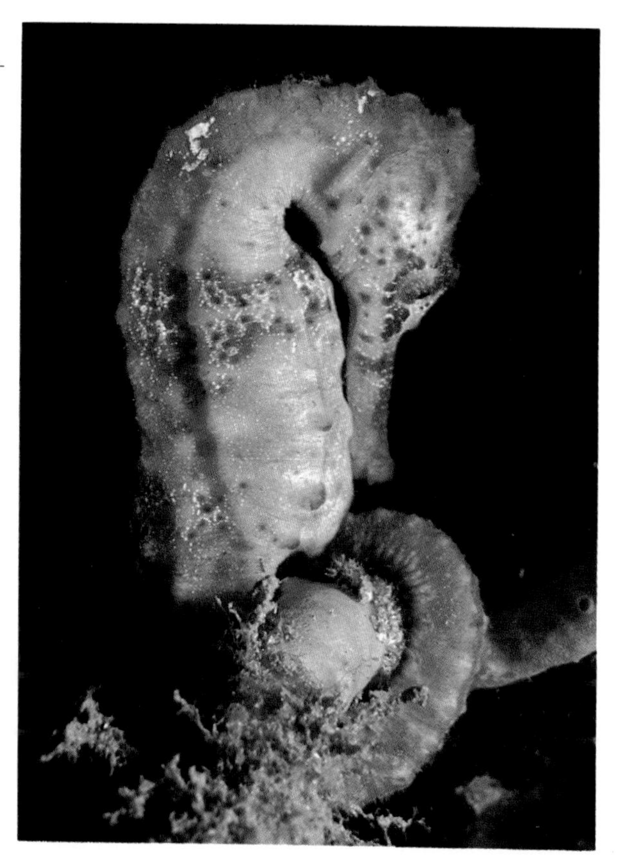

*The seahorses of the Frederiksted pier are justly famous. The brightly-colored creatures gather here in almost unbelievable numbers. Photo: B. Nyden.*

The night-time marine life under the pier is so prolific and captivating that you could find that you spend an hour's dive without venturing more than 30 feet (10 meters) from the ladder. Large and beautiful tube-dwelling anemones are easily found, but they are sensitive to light and will retract, so view them using the outer edge of your dive light's beam. The orange corals *(tubestra)* expose their polyps to feed at night, and clusters of them can be found on the pilings. Brittle stars, who shyly hide in sponges during the day, spread themselves over their hosts at night. The seahorses are easier to spot at night, and their numbers seem to be growing after a few years of relative scarcity. Many of the day-time fishes, particularly puffers, parrotfish, and trumpet fish, can be found in their sleeping phase and can be approached closely. The nocturnal sea urchins prowl the pilings and bottom at night, so caution should be taken not to bump into one. On that note, it would be wise to wear a full wetsuit or jeans and sweatshirt, as it's common to accidentally bump into one of the closely spaced pilings and possibly receive a nasty sting, scrape, or jab.

| | | |
|---|---|---|
| **Typical depth range** | : | 40–150 feet (12–45 meters) |
| **Typical current conditions** | : | None to light |
| **Expertise required** | : | Intermediate |
| **Access** | : | Boat |

Salt River Canyon's eastern slope is another of my favorites because of its reliable visibility (usually 60 feet, 18 meters, or more) and the abundance of schooling fish. Unlike the west wall, the eastern side has a gradual slope from the apex of the canyon (near the shallow reef that crosses the mount of Salt River Lagoon) to the point where it bends towards Buck Island and parallels the north shore. The slope is composed predominately of coral rubble close to shore, but the contours gradually steepen and at the mouth of the canyon there are some beautiful coral-encrusted outcroppings in an

*Salt River Canyon is a gorge that runs into the sea from the shallow, wide-lobed Salt River Bay. At night, coral polyps along the eastern wall spread their tentacles in search of food. Photo: S. Bower.*

*The Pinnacle on the eastern wall is the site of some particularly beautiful coral growth and huge stands of deepwater gorgonians. Photo: B. Nyden.*

area often referred to as the Pinnacle. Gorgonians flourish along the entire length of the eastern wall, and there are some beautiful stands of the deepwater variety (often mistaken for black coral) around 60 or 70 feet (20 meters) near the Pinnacle.

**Marine Life.** One of the most notable features of this dive site is its abundance of fish. Hundreds of creole wrasse hover over the canyon rim during the day, and black durgeons, blue runners, and blue chromis don't seem to mind mixing company. Perhaps this is the reason this spot is popular with the barracuda too. Solitary individuals can be seen waiting motionless for their dinner to come swimming by. The potential dinner targets are so plentiful that a striking barracuda need not wait long; its charge will send the surrounding fish scurrying in all directions like fragments of an exploding artillery shell. Although these predators are territorial and extremely reluctant to back off when first approached by divers, they will eventually give you the right-of-way without baring fangs.

***Hydrolab Project***

Hydrolab is a four-man underwater habitat owned by the National Oceanic and Atmospheric Administration (NOAA) and operated by the West Indies Laboratory on St. Croix. The habitat is located in 50 feet (15 meters) of water in the apex of Salt River Submarine Canyon, on the island's north shore. Since its arrival on St. Croix in 1978, over 70 one-week scientific missions have been conducted with over 200 scientists participating in research projects of national interest. Saturation diving provides long uninterrupted bottom times, which enable the scientists to conduct research they ordinarily wouldn't be able to do if limited to surface decompression requirements.

*From the outside, the Hydrolab appears large. Actually it measures just 16 feet long (5 meters) by 8 feet (3 meters) in diameter. Photo: S. Blount.*

*Scientists living inside the Hydrolab are able to continuously observe the reef outside through the lab's large port. Photo: S. Blount.*

## Salt River Canyon (West Wall) 4

| | | |
|---|---|---|
| **Typical depth range** | : | 40–130 feet (12–39 meters) |
| **Typical current conditions** | : | Usually light, but can exceed 2 knots when high swells are present |
| **Expertise required** | : | Intermediate to Advanced, depending on weather |
| **Access** | : | Boat |

Salt River Submarine Canyon is one of the most popular diving areas on the island of St. Croix. Two distinctively different wall dives are available within 1500 feet (450 meters) of each other. This description is for the west side of the canyon just before the wall takes a bend towards Frederiksted and roughly parallels the north shore of the island. At this point the wall begins at 40 feet (12 meters) and is nearly vertical to a depth of 150 feet (45 meters) where it meets the sandy canyon floor. Average visibility is around 60 feet (21 meters) but can be drastically reduced during high sea conditions. Heavy seas pounding over the shallow reef at Salt River Lagoon create a hydrostatic head which exits out the canyon along the west wall. The inner lagoon water is very murky and the outflow generally reduces water visibility on the west wall to less than 30 feet (10 meters).

Once you've geared up and rolled into the water you can generally spot the rim of the wall from the surface. As you glide down over the edge you'll notice that the patchy stands of elkhorn coral are replaced by plate corals as the predominant species.

*Stunning deepwater gorgonians hanging from the vertical surface of Salt River's west wall are often mistaken for black coral. Photo: S. Bower.*

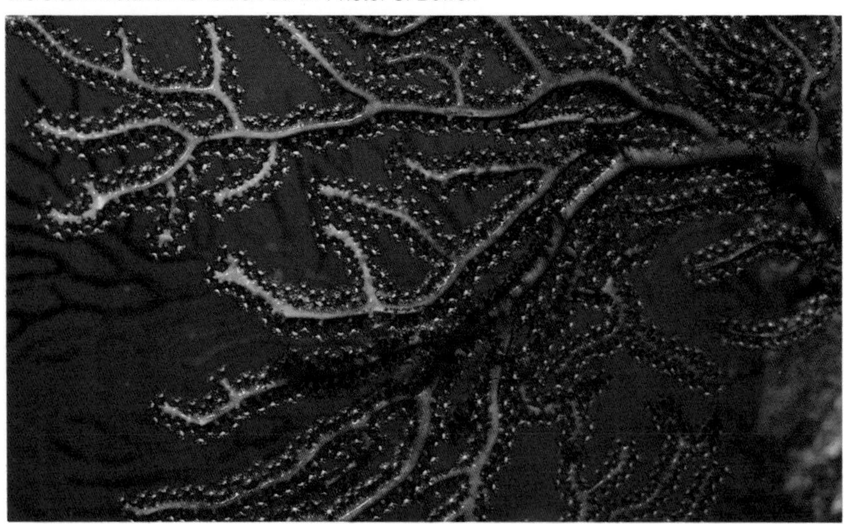

*Large purple tube sponges stud the crevices and rock faces that run the length of the west wall. Photo: B. Nyden.*

### Flashlight Fish

St. Croix's northern shore is home for one of nature's most unusual fish, the flashlight fish *(Kryptophaneron alfredi)*. These fish possess bioluminescent light organs under each eye that can be turned on and off with an eyelid-like visor that rolls up from below the eye to cover the organ. The light is apparently used for feeding and navigation, and is produced by symbiotic bacteria that are cultured within the specialized organs. Flashlight fish are generally found in very deep water, but they have been observed at depths as shallow as 100 feet (30 meters) during the new moon phase. In order to spot these elusive creatures, dives must be made in total darkness.

## Northstar Beach* 5

| | | |
|---|---|---|
| **Typical depth range** | : | 5–50 feet (2–15 meters) to the wall, 50 feet (15 meters) to well below 130 feet (40 meters) over the wall |
| **Typical current conditions** | : | None to light inshore, light to moderate on the wall |
| **Expertise required** | : | Intermediate |
| **Access** | : | Beach or boat |

Off Northstar Beach is a slightly more distant, less frequently dived drop-off which shares some similarities with Cane Bay, but is clearly a different site. In common, there is a forereef, gently sloping to depth of 50 feet (15 meters) at distance of 600 feet (180 meters) from the shore where the wall begins its descent to 300 feet (90 meters). The site is visited by dive boats from Christiansted, or it can easily be dived from shore.

*Plentiful subjects for wide angle photos, such as this anchor, can be found at Northstar.*
*Photo: B. Nyden.*

*Urn-like basket sponges provide a visual counterpoint to the thickets of staghorn coral at Northstar. Photo: B. Nyden.*

To get to the best entry area, go past Cane Bay where the blacktop road turns to dirt and continue west, 0.7 miles (1 km) to a small turn-out off the road to your right. The entry here is over coral ledge, and is home for many sea urchins, so some caution is required, particularly when there is any surf. As you swim out from shore, head slightly west (to your left), bearing approximately 340°, so that when you reach the drop-off you will be directly out from the lone white house on the beach. The swim out will take you over forests of elkhorn and staghorn coral, sea fans, sea rods, and other soft corals, stoplight parrotfish, butterfly, and damselfish in increasingly deep water until you are over a 10 foot (3 meter) indentation in the wall face. At 60 feet (18 meters), this spot can be identified by a large anchor sitting on the sand shelf. Toward shore from this anchor is a good-sized undercut cavern filled with a large school of silversides that seem to dance in the water. From this spot the wall can be dived in either direction, providing interesting formations of coral architecture with small caverns and overhanging ledges. If there is any current running, be sure to swim into it, as this will make the return to this spot and the long swim to shore much easier.

## West End Reef: Butler Bay 6

| | | |
|---|---|---|
| **Typical depth range** | : | 10–40 feet (3–12 meters) |
| **Typical current conditions** | : | None to light |
| **Expertise required** | : | Novice |
| **Access** | : | Beach |

The area off Butler Bay is fine dive site, particularly if you are planning a day on the west end of St. Croix. You could easily visit this site between a mid-day and a night dive at the Frederiksted pier. Butler Bay is about two miles north of Frederiksted (there's only one road, which parallels the coast) past Rainbow and Sprat Hall beaches. Both of these are fine sunning and swimming spots, and the latter has a lunch-time restaurant. When the road makes a sharp U, you have found Butler Bay. It can also be identified by the ruins of a concrete dock in the middle of the bay. The beach is of moderate size and sometimes a bit crowded, but it's still an excellent spot for someone who is staying ashore while you dive.

**Marine Life.** Entry here is very simple: the soft sand bottom just slopes gently out to a depth of about 20 feet (6 meters). Due west from the beach, you will encounter seemingly endless flat plains of seagrass, but don't sell it short and head right for the fish-filled patch reef just to the south, because often these areas can prove fascinating. A bit farther along

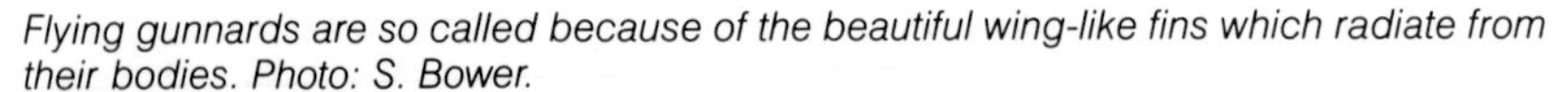

*Flying gunnards are so called because of the beautiful wing-like fins which radiate from their bodies. Photo: S. Bower.*

the flat bottom, a large turtle was slowly sculling out to sea (this was to be one of three turtles seen on this dive). A few more yards, and what appeared to be a log sticking out of the sand was the top of a snake eel. The eel seemed totally undisturbed by my slow approach, and I was able to get very close to him. You can't always be guaranteed to see this much marine life on every dive, but keep looking, because it is there. Although most of the conch on these plains have been fished out, you'll usually spot a few creeping along the sand, and according to those who dive here a lot, it's not uncommon to see dolphins passing through.

If these flat seagrass plains sound too boring, then swim out from the beach heading southwest; about 300 feet (90 meters) out you will find some patch reefs coming out to the sand bottom. Most of the large ones are in 25 feet (8 meters) of water, and they are teeming with small tropical fish. These reefs must have one of the densest populations of fishes in the Virgin Islands. There are brown and blue chromis, schools of French grunts, a variety of wrasses, and squirrelfish, damselfish, and basslets, all swimming around the massive boulder, brain, and star corals. Christmas tree worms adorn the surface of the corals, and tiny cleaner shrimp wiggle their antennae to lure a host fish in for a cleaning.

As the depths here are shallow (you can find 50 feet, or 15 meters, if you swim out far enough), you can make a long dive by first trying your luck at finding something of interest on the seagrass plain, then turning south and returning via the patch reefs.

*Large marine creatures, such as this sea turtle, are a common sight at Butler Bay. Photo: S. Bower.*

## Cane Bay Drop-off 7

| | | |
|---|---|---|
| **Typical depth range** | : | 5–40 feet (2–12 meters)to the wall, 40 feet (12 meters) to below 130 feet (40 meters) on the wall |
| **Typical current conditions** | : | None to light inshore, light to moderate on the wall |
| **Expertise required** | : | Novice inshore, intermediate on the wall |
| **Access** | : | Beach or boat |

Three dive sites share the same reef structure that runs for four miles along St. Croix's north coast: Cane, Northstar, and Davis Bays. Although the overall characteristics of the dives are similar (shallow inshore zones sloping out through an area of elkhorn coral to the forereef and wall) differences in the access, the marine life, and the underwater geography make these distinctly different dives.

A shore dive to the Cane Bay drop-off begins at the lovely beach of Cane Bay, which is a half-hour drive west of Christiansted. Although the area is busy on weekends, there is seldom a problem parking just off the road next to the beach. Suiting up right out of your car is the best procedure here. Air fills are available from a dive shop just across the road from the beach, or another a mile (1.6 kilometers) east on the road from Christiansted.

*The shallow reef area just in front of the beach at Cane Bay is an excellent way to end an excursion out to the deeper wall area. Photo: S. Bower.*

*The clarity of the water at Cane Bay can be deceiving. It's easy to descend 10 or even 20 feet below your planned depth without noticing. Photo: S. Bower.*

**Marine Life.** There is often some surf here, but look for areas of calm water between the breakers; these indicate a safe passageway through the coral heads. A surface swim to the drop-off is recommended to conserve your air, but it is fairly long—about 450 feet (135 meters)—so pace yourself and enjoy it. You will pass over black sea urchins in their daytime retreats, as well as damselfish, wrasses, sea fans, and small soft corals. Past the elkhorn coral "trees" the sand bottom will slope gradually, with clumps of boulder coral and gorgonia. In this area parrotfish, trumpetfish, sand tilefish, and tobaccofish abound, and often young turtles will be seen. As you continue off shore, schools of black durgons (a type of triggerfish) and the deeper blue of the open sea will tell you that you are near the crest of the drop-off. Descending to 40 feet (12 meters) at this point will put you on one of many sandy plains separated by massive coral heads. Swimming between these heads reveals plate, fire, and brain corals, vase, tube and pipes-of-pan sponges as well as soft corals, blue and brown chromis, sergeant majors, damselfish, fairy basslets, and morays secluded in their lairs. Closer inspection shows christmas tree worms, fan worms, anemones, banded shrimp, and gobies. Out toward the deep blue of ocean may be bar jacks, eagle rays, or a huge school of spadefish often seen gliding over the reef crest. Reports have it that sharks sometimes pass by in the deeper water (100–120 feet, or 30–36 meters), but seeing them is very rare.

## Davis Bay 8

| | | |
|---|---|---|
| **Typical depth range** | : | 20–130 feet (6–40 meters) |
| **Typical current conditions** | : | Generally none, but usually heavy surf |
| **Expertise required** | : | Advanced, (occasionally Intermediate) |
| **Access** | : | Beach |

Davis Bay is located on the northwest shore of St. Croix, approximately 45 minutes by car from Christiansted. This area is not dived as frequently as Cane Bay and Northstar Beach because it is very remote and because surf conditions are generally rougher here. It is, however, the most scenic beach of the three. The picturesque white sand beach is enclosed by a rocky precipice on the west end and extends for several hundred yards to the east. There are plenty of coconut palms to offer shade, and the area is an idyllic setting for beach dive and picnic.

*A fairy basslet nestles up to a sponge at Davis Bay. Although generally rougher, Davis Bay is a spectacular beach dive. Coral formations and other growth abounds along the gentle slope. Photo: B. Nyden.*

*A diver investigates the peculiar species of brown rope sponge that covers some areas of the wall at Davis Bay. Photo: B. Nyden.*

The beach entry can be made easily from any point within 50 yards (45 meters) of the west end of the beach. It is very important that surf conditions be evaluated carefully. A quick snorkel tour of the surf zone and potential diving area should be made prior to donning scuba gear if at all possible. If conditions seem manageable and scuba is appropriate, a short snorkel through the surf will put you in 20 feet (6 meters) of water very quickly. Descend to the bottom and then follow a compass heading due north until you see the gently sloping shelf fall away. The slope increases dramatically at 40–50 feet (12–15 meters), and once you've reached your predetermined maximum depth you can travel east or west, paralleling the contour of the slope. A convenient way of planning the dive is to make note of an underwater landmark and then travel with ½ your air supply in either direction, returning to your starting point for the beach exit.

Davis Bay's wall has a much gentler slope than Cane Bay and Northstar Beach, but it does have the same unlimited depth range. Deep grooves run perpendicular to the shore, scarring the rocky shelf. The wall is covered with plate coral and a type of brown sponge that resembles moose antlers. The fish don't seem to be as abundant as they are at Cane Bay or the east slope of Salt River Canyon, but there are still plenty for fishwatchers and photographers alike.

## Buck Island Scuba Area*  9

| | | |
|---|---|---|
| **Typical depth range** | : | 15–40 feet (5–12 meters) |
| **Typical current conditions** | : | Generally none |
| **Expertise required** | : | Novice |
| **Access** | : | Boat |

Buck Island (not to be confused with St. Thomas's Buck Island) is two miles (3 kilometers) off the northeast side of St. Croix and is one of the few marine U.S. National Parks. A variety of commercial charter trips are offered from Christiansted and Tague Bay. Some are strictly snorkel-and-picnicking trips, while others cater to scuba divers. The eastern end of Buck Island is famous for its underwater snorkeling trail, and because this park requires boat access it is not as heavily trafficked as its St. John counterpart, Trunk Bay. Approximately one dozen permanent moorings at the snorkel trail have helped to minimize coral damage, and scuba diving is not permitted here. Three scuba moorings are placed on around Buck Island's east point inside the fringing reef. They are placed near a cut in the reef that is relatively shallow—15–20 feet or 5–6 meters—at first and then opens out into a beautiful forereef area in approximately 35 feet (11 meters) of water. There are plenty of small coral caves to explore and a

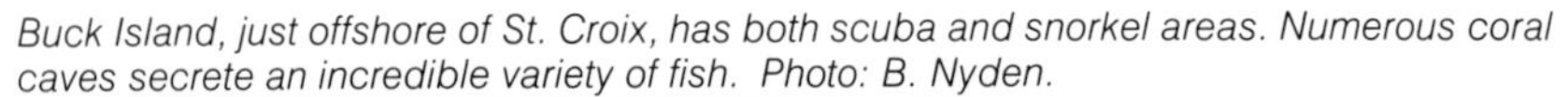

*Buck Island, just offshore of St. Croix, has both scuba and snorkel areas. Numerous coral caves secrete an incredible variety of fish. Photo: B. Nyden.*

*The coral stand, which resembles a wind-whipped desert palm tree, is just one of the unusual formations to be seen at Buck Island. Photo: B. Nyden.*

wide variety of fish. One of the favorites to be found here is the red-lip blenny. This three-inch (one centimeter) fish has a black body with clownish face, red-lips, and what appears to be bushy eyebrows. Red-lip blennies have a peculiar way of sitting outside their holes and jerking about as if they had a chronic cough or were trying to bark and scare away the intruders.

The mooring furthest from the snorkel trail has a much larger opening through the fringing reef and opens out onto a forereef that is predominantly composed of elkhorn coral. Numerous baystack mounds of coral extend to within a few feet (one meter) of the surface. The two scuba areas are separated by just enough distance to make it difficult to cover both openings in one dive. It is also recommended that you stick with a diving tour guide, as these openings are sometimes difficult to find again. It is virtually impossible to snorkel over the reef because it extends to the surface.

*Buck Island is justly famous for its huge elkhorn coral colonies, which thrive in the shallow waters. Photo: S. Blount.*

***Buck Island Reef National Monument***

St. Croix's Buck Island is truly one of the gems of the U.S. Virgin Islands. Its fringing reef and enticing beach attract many visitors, both tourists and native Cruzans alike, so we are fortunate to have the island under the protection of the National Park Service. The island was not always so well favored. The original forest of the eighteenth century was of lignum vitae trees. This hard, dense wood, then used for pulleys, bearings, and ski edges, was harvested and repeatedly burned to allow the island's goat population to graze. The goats gave the island its name; from Dutch *Pocken-Eyland* to *Bocken* to the present Buck. In 1948, the Virgin Islands Government began to administer Buck Island as a park, and in 1961, its supervision was transferred to the National Park Service. Since its protection as a park, the island's natural vegetation has made a bit of a comeback, and it is now well foliated by native species. The endangered brown pelican has taken a liking to the security of the island and now makes it a regular nesting site.

*Scuba is not allowed on the underwater nature trail, which allows snorkelers to enjoy the beauty of the shallow reef while learning about the marine environment. Photo: S. Bower.*

### *The Baths at Virgin Gorda*

A thoroughly enjoyable day trip from either St. Thomas or St. John to the British Virgin island of Virgin Gorda (meaning "Fat Virgin") will let you experience some of the finest and most unusual snorkeling in the Caribbean. Boats leave St. Thomas twice weekly and St. John once a week for the 75-to 90-minute ride to the yacht harbor at Virgin Gorda; for ferry information on St. Thomas, call 774-7920, on St. John, 776-6282. There is an intermediate stop at the British Customs house on the island of Tortola (proof of citizenship is required). From the yacht harbor, which is brimming with luxury sailing craft, cab vans or open jitney busses will shuttle you to a bluff above the Baths. The five minute walk to the sea is a preview of the underwater terrain, as the trail meanders through huge, smooth granite outcroppings. The small clean beach is encircled by these monoliths, which scatter out into and under the clear sea.

Three elements combine to make this such a fine snorkeling site: the remarkable tectonics of the mountainous granite boulders, the extremely good visibility, and the proliferation of marine life. The rocks range from size of a basketball to that of a box car. Some are half underwater, some fully, many create tunnels and passageways lit by shafts of sunlight and large enclosed pools radiant with shimmering fish and sparkling water. The sea here is very clear; visibility averages about 75 feet (23 meters). The fish are plentiful and varied. Attracted by the protection offered by the rocks and by the food carried by the current through the Sir Francis Drake Channel are healthy schools of most of the common Caribbean species. The randomly shaped boulders are covered with every variety of coral and all the little creatures that keep it company.

This site is so special that even the most ardent compressed-air-only diver should hang up his regulator for a day, take the trip to Virgin Gorda, and rediscover the simple pleasures of snorkeling.

*A diver brings up an empty queen conch shell at the Baths. Although isolated on one end of the island, the Baths at Virgin Gorda in the British Virgin Islands are a special treat, and well worth the effort required to reach them. Photo: B. Nyden.* ▶

5

# Safety

**Preparation.** A successful dive trip begins at home no matter where your destination is. There are two areas where a little forethought will help to make your trip more enjoyable and safer too! The first thing to do is to pull out that dive gear, blow off the dust and make sure it still works properly and that no parts are in need of replacement.

The Virgin Islands have some of the finest and friendliest dive guides around. They'll want to make sure that you're adequately prepared to get the most enjoyment out of your trip. Be sure to bring your certification card and logbook and be prepared to do a minimal proficiency check of your scuba skills.

**Hazardous Marine Life.** Diving in the Virgin Islands is relatively safe in terms of hazardous marine life. The bulk of diving injuries are due to coral scrapes, fire coral stings, and punctures from the long spined sea urchin. Coral scrapes should be scrubbed with soap and water and swabbed with antiseptic, as they can easily become infected. Fire coral stings generally take care of themselves, but the application of triamcinelone cream can be beneficial. Urchin spines are generally painful for only a few days; they are quickly absorbed by the body. If you receive a massive dose of spines you should probably have the wound looked at by a doctor to be sure infection does not set in. Long pants will afford some protection against scrapes and stings but the best prevention is just to look where you're going and give the nasty creatures a wide berth.

Jellyfish are uncommon, but there are a few species to be particularly cautious of. Although extremely rare, Portuguese man-of-war (not a true jellyfish) have periodically washed up on beaches. The *cubo medusa* (also called box jellies or sea wasps) also pack a wallop and can sometimes be found in calm areas such as Frederiksted Pier. These animals have tentacles bundled in four clusters. The most dangerous box jelly has a bell 2–3″ (1 centimeter) long with tentacles 1–2 feet (30–60 centimeters) in length. Stings have been known to cause respiratory distress in hypersensitive individuals. If this should occur, be prepared to give mouth-to-mouth (or CPR) and seek medical help.

*Most of the colorful reef inhabitants found throughout the Caribbean can be seen at one or another of the Virgin Islands dive areas. Photo: S. Bower.* ▶

**Diving Accidents.** In the event you or your diving partner are involved in a diving accident (air embolism or decompression sickness), first check to see which decompression facility is available. The decompression chamber at Roosevelt Roads Naval Station in Puerto Rico should be contacted first. If they are not operational try the Hydrolab Chamber on St. Croix as a last resort. Hydrolab's chamber is for support of saturation

Emergency phone numbers
(area code 809 for Puerto Rico and the Virgin Islands):

| | |
|---|---|
| St. Croix Hospital | 773-1212 |
| St. John (Morris de Castro) Clinic | 776-6222 |
| St. Thomas Hospital | 774-1212 |
| USN Seal Team Decompression Chamber | 1-865-2000<br>ext. 4721 or 5291 |
| Hydrolab Decompression Chamber | 778-1608 |
| Diving Accident Network | 1-919-684-8111 |
| U.S. Coast Guard Rescue Coordination | 722-2943 |
| Sunaire Lear Jet Service | 778-0727<br>778-2582 |
| Eastern Caribbean Airways (STT) | 774-5868 |
| Eastern Caribbean Airways (STX) | 778-0630 |

diving missions, is frequently unavailable, and is not permanently located in the Virgin Islands. If both of these chambers are unavailable, call the DAN (Diver Accident Network) number for the nearest operational chamber. If the Roosevelt Roads chamber is up, air transportation can be arranged through the Coast Guard at San Juan.

In the event you need emergency air evacuation to the U.S. mainland, Sunaire has Lear Jet service and Eastern Caribbean Airways (ECA) has a jet-prop service. Both have medical personnel on standby and both are stationed at the Alexander Hamilton Airport on St. Croix.

**DAN.** The Divers Alert Network (DAN), a membership association of individuals and organizations sharing a common interest in diving safety operates a **24 hour national hotline, (919) 684-8111** (collect calls are accepted in an emergency). DAN does not directly provide medical care, however they to provide advice on early treatment, evacuation and hyperbaric treatment of diving related injuries. Additionally, DAN provides diving safety information to members to help prevent accidents. Membership is $10 a year, offering: the DAN *Underwater Diving Accident Manual,* describing symptoms and first aid for the major diving related injuries, emergency room physician guidelines for drugs and i.v. fluids; a membership card listing diving related symptoms on one side and DAN's emergency and non emergency phone numbers on the other; 1 tank decal and 3 small equipment decals with DAN's logo and emergency number; and a newsletter, "Alert Diver" describes diving medicine and safety information in layman's language with articles for professionals, case histories, and medical questions related to diving. Special memberships for dive stores, dive clubs, and corporations are also available. The DAN Manual can be purchased for $4 from the Administrative Coordinator, National Diving Alert Network, Duke University Medical Center, Box 3823, Durham, NC 27710.

DAN divides the U.S. into 7 regions, each coordinated by a specialist in diving medicine who has access to the skilled hyperbaric chambers in his region. Non emergency or information calls are connected to the DAN office and information number, (919) 684-2948. This number can be dialed direct, between 9 a.m. and 5 p.m. Monday-Friday Eastern Standard time. Divers should not call DAN for general information on chamber locations. Chamber status changes frequently making this kind of information dangerous if obsolete at the time of an emergency. Instead, divers should contact DAN as soon as a diving emergency is suspected. All divers should have comprehensive medical insurance and check to make sure that hyperbaric treatment and air ambulance services are covered internationally.

Diving is a safe sport and there are very few accidents compared to the number of divers and number of dives made each year. But when the infrequent injury does occur, DAN is ready to help. DAN, originally 100% federally funded, is now largely supported by the diving public. Membership in DAN or purchase of DAN manuals or decals provides divers with useful safety information and provides DAN with necessary operating funds. Donations to DAN are tax deductible as DAN is a legal non-profit public service organization.

# Appendix I

## U.S. Virgin Islands: Information and Services

The islands are a United States Territory; therefore no passport, visa, or innoculations are needed.

The U.S. dollar is the monetary unit.

The time zone is Atlantic Standard, which is the same as that of the eastern U.S. when the mainland is on Daylight Savings Time. The Virgin Islands are one hour earlier during the winter, when the mainland is on Standard Time.

The U.S. Postal Service handles the mail, so first-class letters are 20¢, postal cards 13¢.

Local (anywhere between the three U.S. Virgins) phone calls are 25¢, and long distance rates to the mainland are surprisingly low.

Some U.S. banks have branches in the V.I., but don't expect to be able to cash your check even if you have an account at the same bank back home. Hours are approximately the same as on the mainland.

Major credit cards are accepted most, but not all, places (particularly few on St. John).

To rent a car you need a U.S. driver's license, a credit card, and usually a reservation. Remember: *drive on the left.*

Electricity is the same as on the mainland (110-120 volt/60 Hz) and no plug adaptors are needed.

Fresh water is at a premium on the islands, so be frugal, and try not to waste it.

The U.S. Virgin Islands Tourist Board will, upon request, send you the following informative brochures:

U.S. Virgin Islands Fact Finder
U.S. Virgin Islands Things To See & Do (St. Croix and St. Thomas/St. John)
Sports
Hotel Rates

### *Dive Operations: St. John*

**Scuba Ventures, Ltd.**
Caneel Bay Plantation
St. John, USVI 00830
(809) 776-6111

**St. John Watersports**
Box 70, Cruz Bay
St. John, USVI 00830
(809) 776-6256

**The Dock Shop**
Cruz Bay
St. John, USVI 00830
(809) 776-6338

### *Dive Operations: St. Croix*

**Christiansted:**
V.I. Divers, Ltd.
Pan Am Pavilion, Christiansted
St. Croix, USVI 00820
(809) 773-6045
**Caribbean Sea Adventures**
Box 3015
Christiansted, St. Croix, USVI 00820
(809) 773-6011
Located downtown, King Christian Hotel, Buccaneer Hotel, and Grapetree Bay Hotel
**Fredericksted:**
Above and Below
12 Strand St., Fredericksted
St. Croix, USVI 00840
(809) 772-3701

*Dive Operations: St. Thomas*

**St. Thomas Diving Club**
Box 4976
St. Thomas, USVI 00801
(809) 774-1376
(Operates at Villa Olga, Pineapple Beach, And Bolongo Bay)

**Watersports Center**
Box 2432
St. Thomas, USVI 00801
(809) 775-0755
775-5649 (Evening)
(Operates at Point Pleasant, Frenchman's Reef, and Sapphire Bay)

**Caribbean Divers**
Box 93, Red Hook
St. Thomas, USVI 00802
(809) 775-1935

**Virgin Islands Diving Schools**
Box 9707
St. Thomas, USVI 00801
(809) 774-8687 774-7368

**Underwater Safaris**
Box 8469
St. Thomas, USVI 00801
(809) 774-1350

**Aqua Action, Inc.**
Box 7576
St. Thomas, USVI 00801
(809) 775-3275 (Day) 775-5894 (evening)
Secret Harbour Beach Hotel

**Joe Vogel Diving Co.**
Box 7322
St. Thomas, USVI 00801
(809) 774-2321
12 B Mandal Rd. RT. 42

**Joe Vogel Diving Co.**
Box 6637, Charlotte Amalie
St. Thomas, USVI 00801
(809) 775-4320

**Chris Sawyer Diving Center (Compass point)**
775-5424 775-7320

# Appendix 2

## Further Reading


Kaplan, Eugene H. 1982. *A Field Guide to Coral Reefs.* Boston: Houghton Mifflin

Greenberg, I and J. 1977. *Waterproof Guide to Corals and Fishes.* Miami: Seahawk Press (Printed on plasticised paper, can be taken in the water. Also available printed on paper.)

Colin, P. 1978. *Caribbean Reef Invertebrates and Plants.* Neptune City, N.J.: T.F.H. Publishing Co.

Niesen, Thomas M., 1982. *The Marine Biology Coloring Book* New York: Harper & Row (Barnes & Noble Books)

Humfrey, Michael 1975. *Sea Shells of the West Indies.* London: Collins

Barnes, Robert D. 1980. *Invertebrate Zoology.* Philadelphia, Penn.: Saunders College/Holt, Rinehart and Winston

Meinkoth, Norman A. 1981. *The Audubon Society Field Guide to North American Seashore Creatures.* New York: Knopf

Travel Information:

Scott, Simon 1982. *The Cruising Guide To The Virgin Islands.* Ft. Lauderdale: BVI Bareboat Association

Diving Information:

Marler, George & Luana 1978. *The Royal Mail Steamer Rhone.* Marler Publications, Ltd.

# Index


Anemones, 67, 79
Angelfish, 43, **43**, 53, 55, **63**
Armando's Reef, 56-57

Barracuda, 69
Basslets, fairy, 57, 77, 79, **80**
Batfish, **29**, **64**, 65
Bigeyes, 57
Blenny, red-lip, 83
Bottle diving, 64
Buck Island, **14**, **19**
Buck Island Reef National Monument, 84
Buck Island Scuba Area, 82-87
Burrfish, 65
Butler Bay, 75-76
Butterfly fish, 75

Cane Bay Drop-off, **12**, 78-79
Capella Island, **16**, 42-43
*Cartenser Senior* Wreck, 42, 54-55
Carval Rock, 34-37, **37**
Christiansted, 16-17, **17**
Christmas tree worms, 65, 77, 79
Chromis, 43, 57, 69, 77, 79
Cimmaron Bay, **21**
Conch, queen, **87**
Congo Bay, 38-41
Coral, **38**, **39**, 43, 55, 65, 68-69, **69**, 75, 79, **83**
- boulder, 77, 79
- brain, **7**, 77, 79
- cup (*tubeastrea*), 40, 51, **54**, 67
- elkhorn, **7**, **31**, 50, 75, 78, 83, **84**
- fire, 37, **42**, 53, **56**, 79
- pillar, **16**, **27**, 43, 53
- pink, **47**
- plate, 72, 81
- staghorn, 53, 75, **75**
- star, **22**, **26**, **59**, 77
- stylaster, 51

Coral caves, 46-47, 50, 82, **82**
Coral polyps, 40, **65**, 67, **68**
Coral World, 15
Cow and Calf Rocks, 46-47
Crabs, 65

Damselfish, 75, 77, 79
Davis Bay, 80-81
Decompression chamber, 90
Dive operations, 92-93
Divers Alert Network (DAN), 90, 91
Dolphins, 77
Durgeons, black, 69

Eels, 65, 77, 79

Fan worms, 65, 79
Feather worms, **22**, 65
Filefish, 52
Fire worms, 65
Flashlight fish (*Kryptophaneron alfredi*), 73
Frederiksted, 17
Frederiksted Pier (Day), 62-65
Frederiksted Pier (Night), 66-67
French Cap—Cathedral, 48-51, **50**
French Cap Pinnacle, 44-45, **44**
Frogfish, long-lure, 65

Goatfish, 65
Gobies, 79
Gorgonians, **26**, **38**, 40, 56, 69, **69**, **72**, 79
- *Pseudopterogorgia*, 36, **36**

Grunts, 57, 65, 77
Gunnards, 65, **76**

Hazardous marine life, 88
Hydrolab Project (NOAA), 70, **70**, **71**, 90

Jacks, 47, 51, 79
Jellyfish (*cubo medusae*), 66

Lameshure Bay, **28**, 30
Long Reef, **25**

Night diving, 40, 52, 65, 66-67
Nocturnal fish, 57, **57**, 65
Northstar Beach, 74-75

Octopus, **66**

Parrotfish, 55, 57, 67, 75, 79
Photograph tips, 37, 41, 43, 45, 47, 57
Pinnacle at French Cap Cay, 44-45
Pinnacle at Salt River Canyon, 68-71
Puffers, 67

Rating systems, 6, 32, 60
Rays
- eagle, 40, 45, **45**, 79
- manta, 34, 37, 40

Runners, blue, 69

Saba Island (Southern Reef), 52-53
Safety, 88-91
St. Croix, 16-19
- beaches, 18-19
- dive sites, 55-87
- lodgings, 19
- recreation, 19
- sightseeing, 18

St. John, 20-23, 30
- dive sites, 30-58
- entertainment, 21-22
- shopping, 21
- sightseeing, 23

St. Thomas, 12-15, **13**, 30
- dining, 15
- dive sites, 30-58
- hotels, 15
- shopping, 12-13
- sightseeing, 13
- transportation, 13-14

Salt River Canyon (East Slope), 68-71
Salt River Canyon (West Wall), 72-73
Scuba diving, 33, 82-87
Sea fans, 75, 79
Seahorses, 65, 67, **67**
Sea rods, 75
Sea urchins, 65, 67, 75
Sea wasps, 66
Sennet, 51
Sergeant majors, 55, **55**, 57, 65, 79
Sharks, nurse, **46**, 47
Shrimp, banded, 79
Silversides, 35, **35**, 41, 75
Sir Francis Drake Channel, **9**
Snappers, 55, 65
Soldierfish, blackbar, **26**
Spadefish, 79
Sponges, **40**, 51, 55, **56**, 65, **66**
- basket, **25**, **48**, **49**, **59**, **75**
- pipes-of-pan, 79
- rope, 81, **81**
- tube, **26**, **41**, **45**, 51, **63**, 65, **73**, 79
- vase, **42**, 79

Squirrelfish, 57, **57**, 77
Star fish, **28**, 35, **36**, 65, 67
Stonefish, 65

Tarpon, **35**, 37, 40
Tobaccofish, 79
Trumpetfish, 65, 67, 79
Trunkfish, 65
Tube worms, 53, **65**
Turtles, sea, 77, **77**, 79

Virgin Gorda Baths, 86, **87**
Virgin Islands
- diving conditions, 28
- diving operations, 24
- general information, 92
- geography, 11, 24
- history, 11
- marine life, 28
- visibility, 27
- water conditions, 26, 27, 29
- weather, 26

West End Reef: Butler Bay, 75-76
Wrasse, 52, 69, 77, 79
Wreck diving, 54-55, 74, **75**

Yellowtails, 55